NEW DIRECTIONS FOR INSTITUTIONAL RESEARCH

J. Fredericks Volkwein, *Penn State University*
EDITOR-IN-CHIEF

Using Geographic Information Systems in Institutional Research

Daniel Teodorescu
Emory University

EDITOR

Number 120, Winter 2003

JOSSEY-BASS
San Francisco

USING GEOGRAPHIC INFORMATION SYSTEMS IN INSTITUTIONAL RESEARCH
Daniel Teodorescu (ed.)
New Directions for Institutional Research, no. 120
J. Fredericks Volkwein, Editor-in-Chief

New Directions for Institutional Research is indexed in *College Student Personnel Abstracts, Contents Pages in Education,* and *Current Index to Journals in Education* (ERIC).

Microfilm copies of issues and chapters are available in 16mm and 35mm, as well as microfiche in 105mm, through University Microfilms Inc., 300 North Zeeb Road, Ann Arbor, Michigan 48106-1346.

NEW DIRECTIONS FOR INSTITUTIONAL RESEARCH (ISSN 0271-0579, electronic ISSN 1536-075X) is part of The Jossey-Bass Higher and Adult Education Series and is published quarterly by Wiley Subscription Services, Inc., A Wiley Company, at Jossey-Bass, 989 Market Street, San Francisco, California 94103-1741 (publication number USPS 098-830). Periodicals Postage Paid at San Francisco, California, and at additional mailing offices. POSTMASTER: Send address changes to New Directions for Institutional Research, Jossey-Bass, 989 Market Street, San Francisco, California 94103-1741.

SUBSCRIPTIONS cost $80.00 for individuals and $150.00 for institutions, agencies, and libraries.

EDITORIAL CORRESPONDENCE should be sent to J. Fredericks Volkwein, Center for the Study of Higher Education, Penn State University, 400 Rackley Building, University Park, PA 16801-5252.

Photograph of the library by Michael Graves at San Juan Capistrano by Chad Slattery © 1984. All rights reserved.

www.josseybass.com

CONTENTS

Editors' Notes

A geographic information system (GIS) is a constellation of hardware and software that integrates computer graphics with a relational database for the purpose of managing data about geographic locations (Garson and Biggs, 1992). GIS technology has been used for almost three decades now. It has, however, for too long been viewed as the province of specialists rather than a generic tool for social scientists. With the proliferation of affordable, easy-to-use, PC-based GIS packages and improved collection and dissemination of geographic data—notably the U.S. Census Bureau's digitized maps—this is changing.

Over the past decade the use of GIS technology and applications in business and academic research has grown exponentially. Integrating spatially oriented data with digitized mapping, GIS tools facilitate the examination of geographic patterns in data, which would be virtually impossible to uncover with traditional statistical analysis. GIS is used by many industries, among them utilities, businesses, law enforcement, transportation, health care, and agriculture, as well as local, state, and federal governments to solve problems and communicate information. It has facilitated natural resource management, land use planning, demographic research, crime analysis, emergency vehicle dispatch, fleet management, environmental assessment and planning, and geological research.

GIS analysis is particularly valuable to local governments, because almost everything that happens in a public policy context happens also in a geographic one; redistricting boards, transportation planners, planning commissions, and crime task forces all must consider questions of *where,* in addition to the usual ones of *how,* and *why,* and *how much it will cost* (Greene, 2000). Williams (1987) estimated that about 80 percent of the informational needs of local government policy makers are related to geographic location. In higher education administration, GIS use is less obvious but interest in learning this technology is growing. Like city managers, campus administrators have to manage people, facilities, equipment, and data. The spatial framework and data integration capabilities of GIS could potentially benefit all areas of campus operations.

Oftentimes, however, the use of this technology in higher education administration is limited to managing facilities and other physical resources. For instance, through the integration of computer-aided design (CAD) drawings and local databases, many campuses have developed systems that help them manage the planning, leasing, constructing, and maintaining of facilities, as well as developing emergency response plans.

The premise of this volume is that, beyond the obvious use in resources management, geographic information systems are likely to play an ever-larger role in many other areas of institutional research. If GIS is still a largely unrecognized opportunity in the IR field, this will probably change. Questions of *where* become increasingly evident to institutional researchers and other campus administrators as they realize that a large part of their data have a spatial component. In a GIS system, address records can be geocoded to their census block group or tract. Given the many address records colleges and universities have on file, research questions inevitably arise that may be addressed through a GIS system. For instance, institutional researchers might be asked to use GIS tools to assist a development office in planning fundraising campaigns by strategically locating areas with a higher level of alumni giving, to assist admissions officers in targeting recruitment and solicitation efforts, to assist continuing education directors in identifying potential markets for nontraditional students, and to assist academic units in optimizing classroom utilization.

Are we equipped with the skills necessary to embark on such projects? For most of us, the answer is probably no. Traditionally, GIS techniques have never occupied a central place in the methodological instruction of social scientists in general, or institutional researchers in particular. Until recently such skills were the exclusive domain of computer scientists, geographers, and marketing consultants. It is clear, then, that there is a need for generalist knowledge about GIS tools that at least introduces the IR professional to their potential applications, concepts, data, and methods.

The purposes of this volume are first to present institutional researchers with a framework for analysis and decision making through the use of geographic information systems and second to discuss examples of GIS applications that are best suited for use by institutional researchers.

The first chapter offers an overview of the GIS concepts, tools, types and sources of data, and types of analysis. The chapter sets the background for the remainder of the volume, which addresses specific examples related to particular applications of GIS in institutional research.

In Chapter Two, Victor J. Mora discusses how a major public research university has been using GIS technology to inform decisions in the areas of planning and implementation of recruitment strategies and tactics. A major strategic planning issue for college and university admissions is the development of inquiries that facilitate attainment of stated institutional enrollment goals with respect to quality, quantity, and diversity. This is a challenging task in light of the difficulty of identifying potential prospects from a very large population and of the increasing competition for students among institutions. The GIS approach described by Mora links current, detailed demographic data at the block group level with local information on enrollees to identify geodemographic profiles of those potential students who are most likely to enroll at his university.

Building on past GIS work at the same university, Manuel Granados presents in Chapter Three several mapping methods to visualize and analyze student enrollment. Through carefully selected examples, the author stresses the importance of selecting the right unit of analysis when creating, analyzing, and interpreting GIS maps.

In Chapter Four, David R. Blough discusses how GIS can be integrated in one of the most widely known IR methods, survey research. The author suggests that GIS can add value to several stages of survey research: research design, data analysis, and reporting. His examples from surveys of the market for higher education are general enough to apply to other contexts and will be insightful both for clients and practitioners of survey research.

Chapter Five illustrates how GIS can serve as an effective tool for space planning, facilities management, and environmental planning of a university campus. It can reveal the most or least utilized space on campus. In this chapter, Nicolas A. Valcik and Patricia Huesca-Dorantes describe the implementation of an integrated GIS for facilities management at a public research university. The project focused on integrating utilities and spatial and relational information, along with distributing this information over an intranet.

Chapter Six explores critical organizational issues that campus administrators are likely to confront when developing an "enterprise system" (a campuswide GIS). Grant McCormick's assumption is that key benefits of a GIS are to be realized with a system that permeates the enterprise, links divisions, integrates data sources to create new understanding, and creates efficiencies by overcoming territorial boundaries.

In Chapter Seven, Daniel Jardine illustrates the use of GIS mapping in analyzing alumni donating patterns and the potential this analysis holds for developing predictive modeling of alumni giving. By linking alumni data with demographic information supplied by census data, GIS can assist alumni offices in gaining greater understanding of where alumni are located and where to focus future campaign activity.

The final chapter draws on the examples of the preceding chapters to highlight the key benefits GIS can bring to institutional research. It also discusses a few resources for those who want to learn more about this technology. The discussion includes information on software, data sources, and training options.

This collection of applications is not intended to turn out full-fledged GIS analysts but rather to expose professionals working in institutional research to this powerful presentation and analysis tool. It is the editor's hope that, by illustrating the potential of GIS with examples of applications from several higher education institutions, this volume will spark the reader's imagination and encourage new approaches to working with this technology in institutional research. Through its clear and intuitive presentation of complex data, GIS has the potential to become one of the most useful tools available to institutional researchers.

Daniel Teodorescu
Editor

References

Garson G. D., and Biggs, R. S. *Analytic Mapping and Geographic Databases.* (Sage University Paper Series on Quantitative Applications in the Social Sciences, series no. 07–87.) Thousand Oaks, Calif.: Sage, 1992.

Greene, R. W. *GIS in Public Policy: Using Geographic Information for More Effective Government.* Redlands, Calif.: ESRI Press, 2000.

Williams, R. E. "Selling a Geographic Information System to Government Policy Makers." *URISA,* 1987, *3,* 150–156.

DANIEL TEODORESCU is director of institutional research at Emory University in Atlanta.

This chapter introduces key concepts, tools, data sources, and types of analysis used in geographic information systems.

An Introduction to GIS: Concepts, Tools, Data Sources, and Types of Analysis

Daniel D. Jardine, Daniel Teodorescu

The use of geographic information systems (GIS) has seen a steady increase since the first PC-based GIS software was developed in the latter half of the 1980s. Prior to that, GIS was run on mainframe computers and was a relatively crude and arcane technology by today's standards. It certainly was not available to the masses as it is today. In the early 1990s, two desktop GIS packages emerged as the leaders: ArcView from ESRI and MapInfo from MapInfo Corporation. Both packages offer all the basic GIS tools and capabilities and are relatively inexpensive and easy to learn. ArcView appears to have been upgrading and improving its GIS package at a faster pace than MapInfo in recent years and has more modules and components to offer at this time.

Initially, the primary uses of PC-based GIS were in business and government. Businesses used the software to map demographic attributes of market areas for analyses such as site selection, marketing campaigns, and so forth. Customer databases would then be overlaid to analyze the geographic extent of their customer base; the practice was commonly referred to as market penetration analysis. Locations of "competitors" would also be mapped out for competitive analysis. Government used GIS technology for mapping and analysis in areas such as zoning, flood plain analysis, and fire and police districts. One example at the federal level involved Housing and Urban Development (HUD) using GIS to map locations of Community Development Block Grant funding projects. Project locations were overlaid on sociodemographic data to determine if funding was being used in areas with the greatest need. The HUD application constitutes an example of program assessment using GIS.

An interesting illustration of how popular GIS technology has become today is its prominent role in the popular television drama "The District," where a Washington, D.C., police chief tracks crime locations on a wall-sized screen using a GIS. Other examples of how GIS technology has made its way into our everyday lives include the variety of Internet sites providing maps and driving directions, and similar technology available in many automobiles.

But what exactly is GIS, and how can it be applied in institutional research (IR)? In this chapter we present key concepts, tools, data sources, and types of analysis used in geographic information systems to answer this question.

GIS Concepts

A phrase many use in referring to GIS is "computer mapping." GIS is about creating maps on a computer for a variety of descriptive and analytical purposes. GIS can help planners and analysts "visualize" data to better understand patterns and concentrations of spatial phenomena. GIS also has the useful ability to portray layers of information, to help uncover spatial relationships among multiple sets of data. A typical GIS "session" involves bringing in various map layers for analysis.

Map layers can take the form of points, lines, or areas (see Figure 1.1). Points represent phenomena that have a specific location, such as homes, businesses, colleges, schools, and crime sites.

Lines represent phenomena that are linear in nature, such as roads, rivers, and water lines. Areas represent phenomena that are bounded (states, counties, zip codes, school districts, census tracts). For example, a higher education institution may want to create a map illustrating the housing locations of off-campus students. A map would typically include (1) the layer of student housing locations represented by points; (2) a map layer portraying streets, represented as lines; and (3) some form of a bounded area layer such as villages or towns, and city wards.

It is important to note that the extent to which one can match data to base maps goes well beyond the familiar examples of mapping state, county, and town data. For example, an excellent use of GIS is in the area of facilities management. Both MapInfo and ArcView have the ability to import AutoCAD drawing files, the most popular format for building and room drawings. Characteristics of each building and room can be associated to the drawings in a GIS. Many higher education institutions have already developed such applications.

Perhaps the most important concept involved in using a GIS is that of associating, or "attaching," attribute data to a spatially referenced *base map*. For example, picture a map of the United States with the state boundaries

Figure 1.1. Points, Lines, and Areas

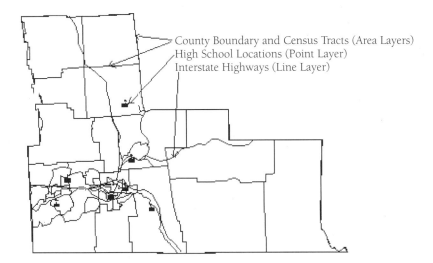

County Boundary and Census Tracts (Area Layers)
High School Locations (Point Layer)
Interstate Highways (Line Layer)

easily visible and distinguishable. This common base map in a GIS would contain the name of each state and, importantly, the coordinates (latitude and longitude) of each state boundary. With this information, a GIS can display a simple base map of the United States by state. A database of socio-economic data such as population, median income, and racial distribution for each state in the country can then be associated or attached to the state boundary map layer. In social sciences research, a GIS may associate the demographic information in the database to the base map by matching the name of each state in the base map to the name of each state in the database. It is this capability of matching up or "merging" data in a database to a base map that is at the foundation of nearly every analysis employing GIS technology.

It is therefore extremely important that the data contain a locational identifier in order to be mapped in a GIS. Typical examples of locational identifiers are street address, zip code, county, state, and census tract. If this information is in the data, then the data can be associated to a base map and portrayed and analyzed in a GIS. The term used to describe the associating of attribute data to a base map in a GIS is *geocoding*, or geographically encoding the data to allow it to be mapped. Address-level data are typically geocoded to a street-level base map, county statistics are geocoded against a county-level base map, and so forth.

Another key concept associated with GIS is that it can be a tremendous reporting tool. One way to think about GIS is that it is a "visual communication tool." Think of a standard data report that lists the number of students by county who attend a particular higher education institution. The counties would be listed in one column and the number of students in an adjacent column. The table could be sorted alphabetically by county or by number of students. This type of report does not have a "spatial" dimension illustrating the location of each respective county. Once these data are geocoded and mapped, one can add a powerful dimension to the communication and "absorption" of the information on the reader's part. Concentrations and patterns immediately come alive. At many institutions in upstate New York, for example, there is a clear upstate-downstate distinction among the student body. A map can illustrate this distinction much more powerfully than a table of county names and numbers. A picture is indeed worth a thousand words.

GIS Tools

Once the data have been geocoded, GIS packages offer many query and analysis tools. For example, if one wanted the GIS to "show" all of the states in the United States with a population of at least four million, one would use the GIS's query capability; it would highlight all of the states meeting that criterion. This simple query would allow one to see the distribution of highly populated states in the nation. Are they concentrated in a particular section of the country, or are they spread out? Are the highly populated states near coastlines, or inland, or both? This is the kind of spatial investigation that GIS provides an analyst.

In the institutional research world, the critical questions may be: "Are we recruiting in areas that have large college-age populations?" "Are we recruiting in areas from where our best students have come in the past?" "Are we mailing our latest alumni donations campaign materials to affluent areas where alumni currently reside?" "Are we attracting more students from the major city nearby or fewer?" "Are we using our buildings and rooms efficiently?" These are just a few of the many questions facing IR analysts every day that could be addressed to some degree through the use of GIS software.

Other tools allow the user to simply click on a data point to access all of the data tied to that point in the original data table. For example, if a database of student information is geocoded and mapped, one simply clicks on a point on the map and retrieves all of the data pertaining to that student in the original data table. The point is then labeled with any field of information from the table. If SAT scores are in the data, a simple query displays the locations of all students scoring above 1,200. Other queries include the location of alumni who gave more than $250 in the last campaign or the location of classrooms on campus that held fewer than two classes last semester. The possibilities are virtually limitless. The advantage of performing such queries in a GIS is the addition of the spatial or visual element, which in many cases enhances the ability to comprehend the information meaningfully.

Another way of thinking about this capability is that GIS can uncover patterns or relationships that otherwise might be buried in tables of data. A map created in a GIS, for example, could show that the majority of large alumni donations come from a wealthy concentration of suburbs near the closest major city, or that the most underused classrooms are in a particular wing of a building, or that a higher percentage of students from county A scored over 1,200 on their SATs than from county B.

Maps produced with a GIS can be printed on color printers or large plotters. They can also be exported to file formats supported by presentation software such as PowerPoint.

Data Sources

There are a variety of sources of GIS data. Both MapInfo and ArcView include map layers, such as county and zip code boundaries with basic demographic data attached, city locations, and so on.

The U.S. Census Bureau is a tremendous resource for base maps and demographic data that can be used in a GIS. State, county, metropolitan statistical area (MSA), zip code, census tract, and census block group are all levels of base maps that are available from the Census Bureau. The U.S. census demographic data can be easily geocoded to these boundaries and include more than three thousand demographic variables relating to statistics on population, housing, income, and education. The Census Bureau has a variety of online tools to make access to geographic data relatively easy. There are also a multitude of Websites with geographic data available, often free of charge.

Many value-added companies enhance existing files from the U.S. Census Bureau. For example, several companies generate projections from census data that have become somewhat dated; census data are updated every ten years. Value-added companies also create special databases that classify areas into certain neighborhood types. Such classification can be particularly useful in recruiting analysis.

GIS has the ability to import data from most popular data management formats: DBF, Excel, Access, and of course fixed-length or delimited ASCII text files.

Types of Data

The type of analysis one wants to conduct with GIS depends largely on how data are measured. The data that are usually mapped in a GIS can be categories, counts or amounts, ratios, or ranks.

A *category* is a group with similar characteristics. For example, an admission office producing a map of areas of recruitment could categorize high schools by type of control, public or private.

Counts and amounts can be used to map discrete features (number of students at each high school within the state) or continuous phenomena (household income by census block).

A *ratio* is used to allow comparison of data between small and large areas and between areas with many features versus those with few. When using counts or amounts to summarize data by area, analysts should be aware that such data types can skew the patterns if the areas vary by size. To avoid false interpretation, GIS analysts can use average, proportion, and density to summarize indicators by area. One might be interested, for instance, in mapping the average number of people per household, or the proportion of high school students in total population by census block. Mapping density allows the analyst to see where features are concentrated; it is particularly useful in displaying distributions when the size of the areas summarized varies greatly. Mapping the population per square mile by census tract is an essential analysis when deciding on the location, for instance, of a future campus.

Rank shows relative value rather than measured value. Rank can be expressed either as text (very satisfied, satisfied, neutral, dissatisfied, very dissatisfied) or numbers (one through five). For example, senior survey data could be mapped to examine whether satisfaction with the college experience is higher for in-state students than for out-of-state students.

It is important to note that to understand the data, GIS analysts often create multiple maps using each of the variable types discussed here. For example, to understand the distribution of Hispanic high school students in a state, one might want to create maps showing total Hispanic population by county, the percentage of Hispanics in the total population, and the density of the Hispanic population.

After determining the type of data to map, the next decision a GIS analyst has to make is whether to map individual values (by assigning a unique symbol) or to group the values into classes. This decision always involves a tradeoff between presenting the data values accurately and generalizing the values to uncover patterns on the map.

As with statistical analysis, it is important to remember that in deciding how to present the information on a map, one should always first consider the purpose of the map and the intended audience. If, for instance, one wants to explore the data to see what patterns and relationships exist in them, the analyst would probably want to display more detail and use various map types. A good start is mapping individual values if one is unfamiliar with the data or area being mapped. The simple display of individual values might also help in deciding later how to group the values into classes.

If one wants to present the map to academic decision makers, however, using classes to group individual values becomes a necessary exercise. Finding patterns and being able to compare areas quickly is especially difficult when the range of values is large. Rank often lends itself to being mapped as individual values; since most Likert scales used in higher education research often involve a maximum of five values, the other numeric data types usually require some kind of aggregation. When mapping ranks with more than eight or nine values, most GIS analysts would recommend grouping them into classes since too many different symbols on a map can make it difficult to distinguish the ranks. Such grouping can be done by simply assigning the same symbol or color to adjacent ranks.

For count, amount, and ratio, grouping individual values in classes is usually recommended for more than twelve unique values. The upper and lower limits for each class can be specified manually or derived by the GIS tool, depending on how the data values are distributed. The grouping schemes most frequently used by GIS software are the equal interval, quartile, and standard deviation.

Usually four or five classes are enough to reveal patterns in the data without confusing the reader. However, if one uses fewer than three or four classes, there might not be much variation between features and therefore no clear patterns will emerge.

Types of Analysis

Mitchell (1999) identifies the most common geographic analysis tasks that can be used to inform decision making: (1) mapping where things are, (2) mapping the most and least, (3) mapping density, (4) finding what's inside, (5) finding what's nearby, and (6) mapping change. We'll discuss only three of these types, since they seem to be most relevant to the type of GIS-related questions institutional researchers typically ask.

Mapping Where Things Are. In this simple analysis, maps are used to identify individual features or to look for patterns in the distribution of features. The purpose of this analysis is to find places that meet a certain set of criteria or to show areas where one needs to take action. For instance, in using GIS a metropolitan university could geocode the addresses of new faculty. Examining where faculty reside could lead to exploring the causes for the patterns revealed by the map. For example, one could see whether the rising median price in the nearby housing market is impeding junior faculty from living close to the campus.

Mapping the Most and Least. Mapping features on the basis of quantity adds an additional level of information beyond simply mapping the location of features. An example of this type of analysis would be mapping the locations of high schools to assist admissions counselors at a selective institution in planning recruiting visits. Mapping the high schools according to the number of students with SAT scores higher than 1,200 would give admissions officials a better picture of where to concentrate their recruiting efforts.

Mapping Change. GIS allows analysts to map changing conditions in a place over time. Knowing what has changed can help higher education planners understand how things evolve over time, anticipate future conditions, or evaluate the results of a past action or policy. Analysts typically want to map how much or how fast a place has changed. Rather than simply mapping the conditions at two times, they usually calculate the difference between the values associated with a feature. The magnitude of the change can be expressed as an amount or percentage and then mapped. For instance, academic planners at a community college might want to analyze changes in the demographic data for census areas in the vicinity of their campus between 1990 and 2000. A trend that shows an increase in the elderly population might prompt the college to add to its programs more personal development courses that are attractive to this age group.

Map Types

Most GIS applications in institutional research are likely to produce *choropleth* maps, which use color patterns or shading to indicate the magnitude of a numeric variable (college-bound population, SAT, household income) or the values of a categorical variable (ethnicity, urban or rural). Figure 1.2 presents an example.

Since it depicts an entire geographic area in a single color pattern, a choropleth map does not show differences in standard deviation of data within an area; therefore it may introduce area bias. That is, marginal changes in the variable being measured cannot be revealed since each geographic unit is assigned to one category or another. This might exaggerate the difference across geographical boundaries.

Isopleth or contour maps are used to create continuous areas that connect like points. For example, areas of wealth and poverty can be shown by connecting census blocks within the same median income range. Isopleth maps eliminate the bias associated with choropleth maps. Rather than suggesting sharp changes in a variable crossing the boundary from one geographic unit to another, they show the actual contour lines of the geographic distribution of a value category. If these contour lines are superimposed on a layer that displays geographic boundaries, the map reader can get a better understanding of the relationship between boundaries and values. This is commonly done in weather maps of frontal systems and topographic maps.

A *cartogram* is a variant of the choropleth map in which the two-dimensional boundaries of geographic units are distorted so that the surface area of each geographic unit is proportional to the amount of the value being measured.

Conclusion

As institutional researchers, we "sit" on a number of data files that can be mapped in a GIS. Student data, alumni data, and admissions and inquiry data all can be mapped in a GIS. Other campus administrators as well have discovered that GIS is an important tool for a variety of tasks. Facilities managers, for instance, have maintained CAD-based maps of their campuses for many years, but now they are adding intelligence to their maps by using

Figure 1.2. Number of Students by "Permanent" County

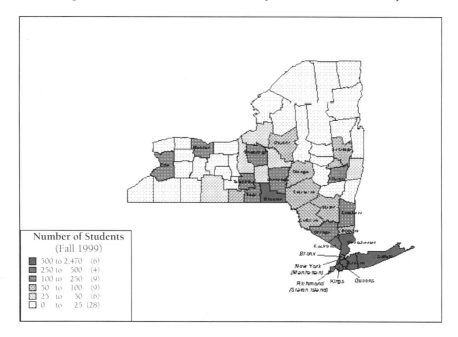

GIS so that they improve their planning and analysis. Admission offices at several universities are using GIS to map out their data on freshman admissions, and institutional advancement officers are mapping their data on alumni. Other administrative branches of a campus that use GIS are campus police, telecommunications, real estate management, grounds management, hazardous waste tracking, and student housing.

Reference

Mitchell, A. *The ESRI Guide to GIS Analysis.* Vol. 1: *Geographic Patterns and Relationships.* Redlands, Calif.: ESRI Press, 1999.

DANIEL D. JARDINE *is a research analyst in the office of institutional research at Binghamton University in Binghamton, N.Y.*

DANIEL TEODORESCU *is director of institutional research at Emory University in Atlanta.*

2

This chapter illustrates how Ohio State University has been using GIS to help with strategic and tactical decisions in student recruitment at its Columbus campus.

Applications of GIS in Admissions and Targeting Recruiting Efforts

Victor J. Mora

In a highly competitive and unpredictable environment, colleges and universities must use technology to develop effective tools to meet demands to improve student recruitment and retention profiles while saving costs. There is increased marketing sophistication in public colleges and more reliance upon technology to conduct computer-based analysis throughout the student recruitment process. One such technology is geographic information systems (GIS). The Ohio State University's Undergraduate Admissions Office began to incorporate the use of GIS into its operations several years ago (Mora, Granados, and Marble, 1997).

Undergraduate Admissions has been challenged to achieve goals set forth by the university administration. These goals involve three general areas: *quantity* (number of students to be matriculated), *quality* (on the basis of academic ability), and *diversity* (race and ethnicity, geographic location, academic area, gender, and so on). The scope of this chapter is limited to the new population of first-year students.

The Ohio State University has been using GIS technology to support decisions in the areas of planning and implementation of recruitment strategies and tactics. At the planning level, GIS has been useful in evaluating current (and selecting potential) geographic markets for new freshman recruitment. At the implementation level, the university has been making decisions on where and how to deploy its limited resources to achieve its recruitment goals and strategies. Research shows that areas most likely to produce the targeted number of enrollees with the desired spectrum of attributes (with respect to quality and diversity, for example) are similar to the

NEW DIRECTIONS FOR INSTITUTIONAL RESEARCH, no. 120, Winter 2003 © Wiley Periodicals, Inc.

areas that have historically yielded greatest returns on outreach and follow-up efforts (Mora and Herries, 1999).

The GIS infrastructure developed at Ohio State is made up of internal and external data sources and tools. The internal source is the undergraduate admissions database, which contains multiyear data on prospective students, on those who inquired, applied, were admitted, paid fees, and enrolled. The external information comes from various sources, the primary ones being Claritas, Inc., the leaders of precision marketing with the use of geodemographics; the Ohio Department of Education, which furnishes enrollment statistics (by grade level, gender, race and ethnicity) for public high schools; national testing services (ACT and SAT), which offer the names and test score information of potential students as well as the addresses of high schools in the continental United States; Geographic Data Technologies (GDT), which makes available geocoding services; and ESRI, which sells GIS software, base map data, and digital mapping software (ArcView, ArcGIS). Other sources constitute tools that are used for developing predictive models, such as data mining software and an in-house software application called Empowering Market Analysis (EMA).

Claritas is an important component of the information structure. It is the primary source of geodemographic databases, which are used for the purpose of market segmentation analysis. The company generates annual demographic estimates and socioeconomic classification of the U.S. population, down to the block group level (which is made up of 350 to 500 households). Claritas breaks down the U.S. population into sixty-two neighborhood types (also called PRIZM lifestyle clusters), on the basis of socioeconomic, lifestyle, and urbanization characteristics. It also assigns each block group (or neighborhood) a predominant PRIZM cluster, under the theory that "birds of a feather flock together." For higher education, this implies that a group of students living in the same neighborhood are likely to share similar behaviors when it comes to higher education decisions. Conversely, this clustering assumes that a student who lives in a rural area does not behave in the same way as one who lives in an urban area (Mora and Herries, 1999).

As has already been mentioned, GIS technology is used to support decisions for planning and implementing recruitment strategies. Here are descriptions of how Ohio State applies GIS at each of these levels.

Planning

The planning process begins with a look at the past. Ohio State evaluates its various market segments by comparing projected and actual outcomes. Each market is analyzed on the basis of patterns of prospective students, applicants, and enrollees, or a given performance criterion such as conversion rate (proportion of inquiries who apply) or yield rate (proportion of admitted students who enroll).

Figure 2.1. The Enrollment Management Funnel

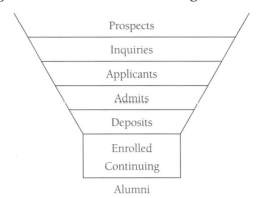

GIS technology has been useful in displaying multiple layers of data on a single map to show information that would otherwise be lost in text, tables, and charts. With these capabilities, the university is able to identify geographic markets for purchasing names of students from testing services and other sources. Purchase of these lists is generally based on test score ranges and potential of the geographic areas, according to past performance or other criteria. These lists make up the annual pool of approximately 150,000 potential students at the beginning of the recruitment process (see the top of the funnel in Figure 2.1).

Because the enrolled profile of new first-year students depends on the profile generated at the beginning of the recruitment funnel, it is critical that the appropriate pool be built right from the start; thus products should be developed that support decisions such as where to purchase names of prospective students, where to develop new geographic markets, where to maintain our efforts, and where to apply market penetration strategies.

Geographic markets are defined and treated differently to meet the needs of the Undergraduate Admissions Office. The Ohio markets, for example, consist of counselor territories, which are sets of contiguous counties. The non-Ohio domestic geographic markets are based on metropolitan statistical areas (MSAs). It is important to note that both the Ohio and non-Ohio domestic markets use the census block group as their geographic infrastructure to conduct segmentation analysis. To conduct such segmentation analysis, GDT geocodes student records by student address; that is, student addresses are assigned to a specific geographic area or to a point location (longitude and latitude coordinates). Student address assignment to a small area, such as the census block group or neighborhood, permits identification of predominant socioeconomic and lifestyle characteristics of the students' neighborhood from the Claritas databases. Once these characteristics are known, profiles of the prospective students can be created and compared with enrollment profiles of previous years. These profiles are then used to identify potential areas where new students can be recruited, or to prioritize high schools that offer the greatest potential to contribute to the university's enrollment goals (Mora, Granados, and Gauchan, 2002).

Analysis of the Ohio market has shown that just thirteen of the sixty-two clusters accounted for 5 percent of all freshman enrollees and that an additional twelve clusters accounted for another 25 percent of new freshman enrollees. The GIS software was used to identify those block groups in Ohio where the twenty-five PRIZM lifestyle clusters are predominant. These geographic areas (neighborhoods) offer the greatest potential for future recruitment because students coming from these areas can be assumed to be more predisposed to come to Ohio State than students from other neighborhoods. Understanding the demographic and lifestyle characteristics of the various student segments makes possible the design of targeted communication strategies. More fine-tuned analyses have been conducted for each territory to understand the impact of the competition on certain student types, or the factors that are important in their college decision-making process as opposed to the impressions they have of the institution that are based on these factors.

This market analysis has been helpful in selecting target student segments and high schools to visit during fall travel. Tables and maps were created to illustrate the type of student coming to Ohio State, the characteristics of their neighborhoods, and the high schools they attended. Outreach and follow-up strategies have been applied from this analysis.

For the non-Ohio domestic market, analyses at the U.S. MSA level have been helpful in identifying the most productive MSAs with respect to new freshmen enrollment. A frequency distribution of all enrollee records by MSA was conducted to identify the most productive MSAs outside the state. Eleven MSAs contributed more than half of the non-Ohio new freshmen over the past several years. Because these MSAs offered the best potential for recruitment, the university has focused on them in generating prospective student lists. These lists are purchased from testing services and other sources. Each targeted MSA is then further analyzed at the block group level to understand the socioeconomic characteristics of the neighborhoods where potential students are concentrated. This information is useful for targeted communication strategies.

Implementation

The university developed EMA, a powerful market analysis tool to support implementation strategies. EMA integrates internal student attribute data with external geodemographic data, digital mapping, and predictive modeling into one interactive and powerful environment.

The power of geodemographic segmentation is placed in the hands of users through this simple yet effective software application. EMA gives users direct access to data and empowers them to conduct analysis and make their own decisions. EMA not only enables users to make strategic and tactical decisions, such as where to target for new prospects, but also helps them determine what type of message to develop for a particular segment of the population.

Segmentation analysis is conducted by using the admission and socio-economic data of prospective students to estimate their probability of enrollment. This analysis is done at the inquiry level and at the admit level. Once this probability is known, it can be used inside EMA as another criterion to identify, classify, or analyze prospective students. For example, it is possible to ask the system to identify all the students with high probability of enrollment, those who live in a specific territory, those who have high ACT scores, and those who are potential candidates for financial aid (depending on estimated household income). Once these students are known, they can be selected to receive differentiated messages that emphasize scholarship or need-based financing (Mora, Granados, and Gauchan, 2002).

Probability of enrollment is assigned to records at two stages of the recruitment process: inquiry and admit. The predictive models are developed using logistic regression, where enrollment is the dependent variable and the student and environment characteristics are the independent ones. Some variables that have shown predictive power are ACT score, campus visit, number of contacts before enrollment, early application or inquiry, distance to the university, income, and college of interest. Having these predictive models developed in-house gives the institution great flexibility and control at lower costs than with use of external providers (Mora, Granados, and Gauchan, 2002).

Here is a process that illustrates, in broad terms, how data are prepared in EMA for the users. First, records of prospective students purchased from various testing sources are sent to GDT for geocoding (assignment of a geographic location). The geocoded records are then merged with geo-demographic attribute data from Claritas and other sources (such as the Department of Education). Data from individual students are entered periodically throughout the recruitment process. Predictive scores are assigned to each student at the inquiry and admitted levels and entered into EMA. These data are used for student follow-up by telecounselors. As student contact information is generated and entered into the admissions database, it is also updated weekly into EMA.

Through the funnel reports functionality, EMA permits users to query, display, and analyze demographic and university enrollment data at various stages of the recruitment process. Funnel reports enable territory managers to have a current picture of their territory, on the basis of multiple segmentation choices, and to compare that picture with data of a year ago. Territory managers can extract records directly from these reports and use student lists for targeted communication.

It is important to point out that, to provide its growing and diverse number of users with current, comprehensive, relevant, and useful information,

EMA is continuously being updated, fine-tuned, and developed. Users of EMA range from strategic recruitment planners, financial aid directors, managers, and evaluators to recruitment supervisors, telecounselors, territory managers, liaisons with academic departments, minority recruitment supervisors, and staff. All these users have their own specific needs for detailed customization, flexibility, and comprehensiveness.

Our office conducts training for users to enable them not only to use the tool but also to understand the data elements within EMA and to resolve problems.

A number of exercises are used in training territory managers on EMA and on analysis of their markets. The first exercise is designed to build skills on the use of EMA's features and functionalities; the second is to use EMA for problem-solving situations.

In the skill-building exercise, use EMA to identify and build a profile of inquiries in your territory *who have applied* to OSU for the autumn quarter. Of this population, determine the conversion rate of those inquiries who said they will apply or probably will apply in the last telecounseling contact. Then identify and build a profile of those *who have not applied* but said they will apply or probably will apply.

In the problem-solving exercise, Ohio State identifies segment X as being very important in your territory; thus you want to increase the number of segment X applications in your territory. A partnership with the Office of Student Financial Aid and a couple of academic departments has been established for outreach and targeted follow-up activities. Several approaches have been suggested. One is to have a campus-visit program for high-ability segment X members who have expressed interest in Ohio State but have not yet applied. The focus of this activity is to educate students about scholarship opportunities. The selection criteria are GPA equal to or greater than 3.5, *or* top 10 percent of the class, *or* ACT score of 28 or better. The other approach is to hold a reception in your territory and invite low-income students who have an ACT (or converted SAT) score of 24 or better. We want to invite these students' parents so they can also learn about merit scholarships and need-based aid from the institution, as well as federal and state grants.

To maximize your results, focus on those students who expressed interest and those who are likely to enroll. The average inquiry predictive score is 10 percent. If you use 30 percent or higher, you are selecting those students who have three times more probability of enrolling than average.

Select a pool of students to invite to campus and also prepare a profile of those students whom you would like to visit. What materials should you bring to each activity, and what approach should be used with each group?

These two training exercises enable users to become proficient in the use of EMA and empower territory managers to understand the data, analyze them, and make decisions. As was previously mentioned, it is critical that users understand the data in EMA; without this understanding, errors can occur, resulting in faulty information and wrong decisions.

Conclusion

Most institutions may begin using GIS by simply mapping their resources and constituencies. Some move on to develop analytic methods for answering specific strategic questions. A few have gone one step further to develop and deploy GIS-based tools for answering more tactical questions such as "Which schools should we visit this year?" or "How should I communicate with different segments of the student population, at specific stages of the recruitment process?" The Ohio State University has been using GIS to help with strategic and tactical decisions in new student recruitment to its Columbus campus. The methodologies and tools that have been developed can be successfully applied in a number of other institutions. Each institution faces its own unique challenges, but the tools and methods employed here are likely to be equally useful in other situations.

References

Mora, V., and Herries, J. "Geographic Information System Technology: An Effective Tool for Improving In-State and Out-of-State Undergraduate Admissions." Paper presented at the Annual Forum of the Association for Institutional Research in Seattle, May 30–June 2, 1999.

Mora, V., Granados, M., and Gauchan, S. "A Demonstration of a Software Application That Integrates GIS Technology with Predictive Modeling in Student Recruitment." Demonstration presented at the Annual Forum of the Association for Institutional Research in Toronto, June 2–5, 2002.

Mora, V., Granados, M., and Marble, D. F. "Applying GIS Technology and Geodemographics to College and University Admissions Planning: Some Results from the Ohio State University." In *Proceedings* of the 1997 ESRI User Conference. Redlands, Calif.: Environmental Systems Research Institute, 1997.

VICTOR J. MORA is associate director of enrollment management at the Ohio State University.

3

This chapter illustrates application of GIS techniques to analysis of student enrollment at a major university.

Mapping Data on Enrolled Students

Manuel Granados

Traditionally, educational institutions have relied heavily on the use of tables and graphs to analyze and present enrollment data. No doubt, a lot of information can be conveyed through a simple table, or a complex one. Graphs can be more effective for presenting such analysis of data, even when they are nondynamic. However, in dealing with information that has a spatial component (such as enrollment data), graphs and tables are often insufficient for communicating the spatial dimension. An effective way to add this extra dimension to the analysis is through maps, since many spatial patterns are easier to identify and understand. This chapter demonstrates how maps can be used to visualize and analyze student enrollment. All of the examples come from the experience accumulated through several years of working in the application of GIS techniques to analysis of student enrollment at a major university.

How Records Are Mapped

The examples given here were created with data from the Office of Undergraduate Admissions and First-Year Experience, the U.S. Census Bureau, the Ohio Department of Education, and Claritas, Inc. To assign student records to a geographic location, they are geocoded; that is, the records are assigned to a census block group depending on their home street address.

Scope, Scales, and Units of Analysis

Mapping enrollment data can be done at a local, regional, or national level. The scope depends on the objectives of the analysis, which also dictate the best basic units to use. Traditionally, states, counties, and zip codes are the most common units of analysis, although they are not always the most appropriate ones.

For an example of a national-level analysis, let's examine the distribution of admits at the Ohio State University for autumn 2002 by state of residence (see Figure 3.1).

This map shows the general distribution of the admitted students; it is clear that more come from the eastern part of the nation than from the western, with Illinois being one of the major contributors.

Now examine the same number of admits, but at the county level (Figure 3.2).

This map allows us to see more detail in the data. For instance, note that not all of Illinois is a contributor of admits; most come from the Chicago area. These two examples illustrate that although the map at the state level offers useful information, the one at the county level gives a little more detail.

To go further into the analysis, look at actual enrollment numbers, shown in Figure 3.3, which shows the count of new first-quarter freshmen (NFQF) at the county level.

Comparing this map with the one that shows admits by county (Figure 3.2), it is possible to see not only which areas produce more enrollments but also which areas have more admits who actually enroll. To fully understand this relationship, it is better to examine the yield, which is the ratio between enrollees and admits. This is depicted in Figure 3.4, where dark areas indicate high yields and light areas correspond to low yields.

Figure 3.1. Admits by State

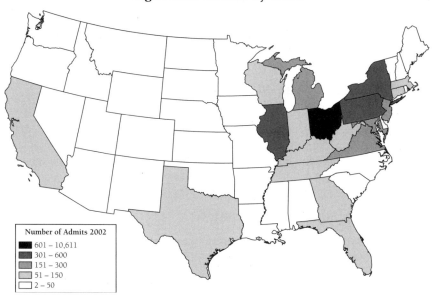

Number of Admits 2002

601 – 10,611
301 – 600
151 – 300
51 – 150
2 – 50

Figure 3.2. Admits by County

Figure 3.3. Enrollees by County (New First-Quarter Freshmen)

Figure 3.4. Yield by County (Enrollments per Admits)

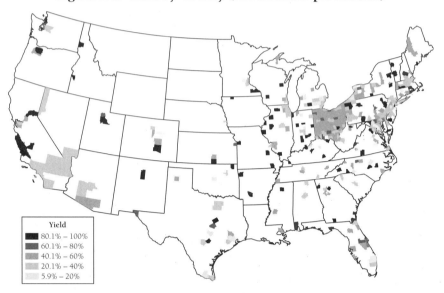

Yield
- 80.1% – 100%
- 60.1% – 80%
- 40.1% – 60%
- 20.1% – 40%
- 5.9% – 20%

Another aspect to consider in mapping student data is determining the appropriate spatial units. Different methods can be used at a local level; one is zip codes, and another employs census block groups. Zip codes are commonly used to map areas and their characteristics such as population demographics. However, they are not the best units for this purpose. As an example, Figure 3.5 shows the median household income at the zip code level for Franklin County, Ohio, and its surrounding area.

The map gives an idea of the spatial distribution of the household income. But how accurate or realistic is that representation? We must remember that a zip code is a mail delivery area created by the U.S. Postal Service to facilitate delivery of mail. Zip codes have nothing to do with the homogeneity of the areas or their socioeconomic characteristics. Therefore we cannot expect a zip code area to have the same characteristics throughout. To describe and understand the socioeconomic characteristics of an area, we have to go to the census block group (BG), which is, according to the Census Bureau, "the lowest level geographic entity for which the Census Bureau tabulates sample data from a decennial census . . . [and which] generally contains between 300 and 3,000 people with an optimum size of 1,500 people" (U.S. Census Bureau, 2002, p. 6).

Figure 3.6 shows the median household income, again for Franklin County and surrounding area, at the block group level.

When comparing this map with the one by zip codes, one sees that the analysis at the block group level offers more detail and may be more realistic. For instance, if we look at zip code 43123, Grove City, the median

Figure 3.5. Franklin County Median Household Income by Zip Code Level

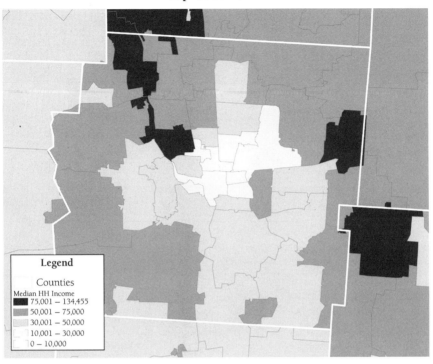

Legend

Counties

Median HH Income

- 75,001 – 134,455
- 50,001 – 75,000
- 30,001 – 50,000
- 10,001 – 30,000
- 0 – 10,000

household income, according to the zip code tabulation, is $58,143. However, analysis of the same area tabulating the median income at the block group level shows the average median income for the twenty-eight block groups involved to be $55,784, with a standard deviation of $14,840 and incomes ranging from $18,229 to $82,117. It is clear that the block group map gives more information than the zip code map. This happens even more in a rural area, such as zip code 44890. For this area, the median household income, according to the zip code, is $36,642, but at the block group level it is $37,086, with a standard deviation of $6,359 and an income range from $26,786 to $46,301. (Note that all the median household incomes are estimates from Claritas, for 2002.)

Next, we go to more detailed information, such as point location. As stated earlier, the process of mapping a student address involves geocoding, which is basically a process of assigning a street address to its corresponding spatial location. In other words, the street address is put in a map. This geocoding can be done at the census block group level or at the latitude and longitude level. The latter is more complex and costly. An alternative to mapping the latitude and longitude of each household is to use the number of households per block group and map that using the population centroids

Figure 3.6. Franklin County Median Household Income by Block Group Level

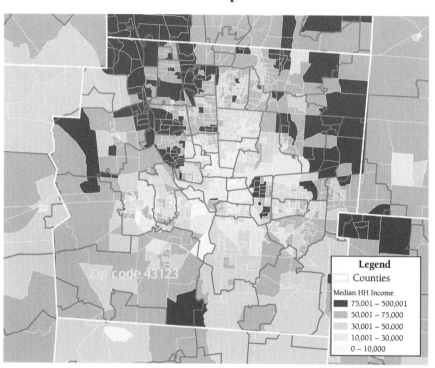

of the block groups. This approach works well at a small scale, as with a metropolitan statistical area or at the state or national level.

Figure 3.7 is an example of latitude-longitude mapping.

The map shows the number of enrollees (NFQF) for autumn 2002 whose houses are located at a precise latitude and longitude, each address being marked by a specific dot. The background of this map depicts urbanization in five categories: rural, town, suburban, urban low-density, and urban high-density. By overlaying these two layers—location of the students' homes and urbanization—it is possible to get an idea of the origin of the students the university is enrolling. Such a map could help a university create targeted messages, or it could be used to evaluate the type of population the university serves. Because of the scale of the map and the size of the dot, each dot could represent a single student or more than one (in this particular case, up to nine). However, because the objective of the map is to give a broad representation of enrollment by urbanization, the number of students by dot is not that important. One way to conduct a deeper analysis is to do a cross-tabulation of the actual number of students by urbanization and run a statistical analysis, for example, a test of independence.

Figure 3.7. Enrollees by Latitude-Longitude and Urbanization

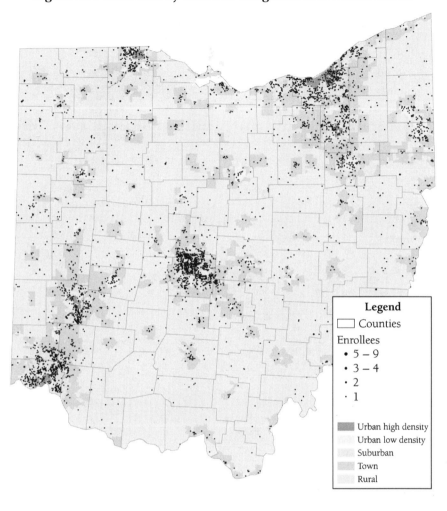

Legend
☐ Counties
Enrollees
• 5 – 9
• 3 – 4
· 2
· 1

Urban high density
Urban low density
Suburban
Town
Rural

Another useful representation of the distribution of enrolled students is to map them at the high school level. Figure 3.8 shows the distribution of high schools in Ohio and their contribution to enrollment for autumn 2002.

Each bubble represents a high school, and its size is proportional to the number of enrollees it yields. One can enhance this representation by selecting a specific type of student in each high school (for example, by academic ability, race or ethnicity, academic interest, and so on). In this case, the high schools shown are those who have the type of student selected. This kind of map can help admissions counselors in planning high school travel.

Instead of selecting a specific type of student, each mapped area—in our case, high school—can be represented by the distribution of the student

Figure 3.8. Enrollees by High School

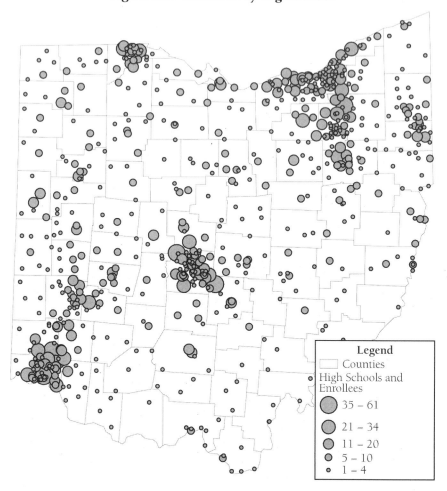

types. Figure 3.9 is an example of this type of representation, where the areas mapped are admissions territories; inside each one a graph represents the local distribution of enrollees according to their estimated median household income.

The size of the pie is proportional to the number of enrollees from each territory. University administrators can use this type of map to help understand the composition of the student population by the territory in which they live.

The preceding examples present some ways to map enrollment and student characteristics by using spatial units and limited data manipulation. What follows is a description of an alternative that requires more complex data analysis.

Figure 3.9. Distribution of Median Household Income by Territory

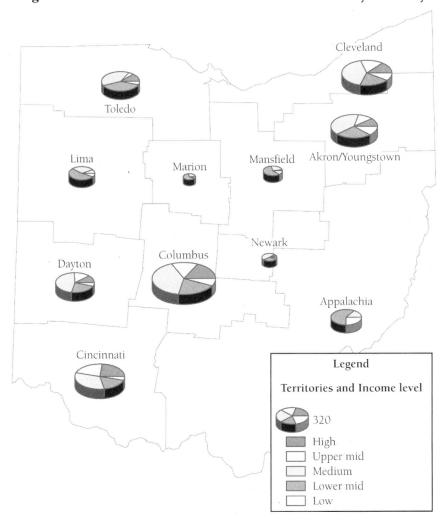

Propensity of Enrollment

One way to analyze propensity of enrollment is to look at the geodemographic component; another is by looking at the specific individual characteristics of the student; and a third way is a combination of both. This section presents an example of the first approach and also offers some comments about the third one.

The goal of assessing the propensity of enrollment of a student population according to its geodemographic characteristics is to identify which groups of students are likely to enroll at a university. To conduct such an analysis, it is necessary first to identify how to group the students and later how to measure that tendency.

One way to group the students is by means of their socioeconomic characteristics. All the students with similar socioeconomic and demographic characteristics will belong to the same group. For this we use the work done by Claritas. This company analyzed the socioeconomic, demographic, and lifestyle characteristics of the U.S. population and in so doing created a set of sixty-two homogeneous groups, called PRIZM clusters. They also assigned each census block group in the United States to a predominant PRIZM cluster. Therefore, if we know the block group where a student lives, we know the most probable PRIZM cluster where he or she belongs. With this information, a PRIZM cluster profile can be created for the whole enrollment population, and it can be compared against the profile of the whole state or region. With this comparison, it is possible to assess which groups have a high propensity of enrollment. An example of using this method is Figure 3.10.

In this case, the PRIZM cluster profile for the enrollees is created and compared against what is considered an "admissible population." That is, the enrollment population is compared not against the whole population, but against the sector of the population that is considered admissible. To identify such admissible population, all the student records available for the state are used. The university receives a large number of inquiries and also purchases test score records of students for the whole state. These two sets, inquiries and purchases, give more than 150,000 student records each year. From these datasets, only those students who have a minimum score (ACT or SAT) are considered as admissible.

To measure the propensity of enrollment, some indices are computed. The first one is the "composition," which is the percentage relative to the total that belongs to a specific group. This is calculated for the enrollment and the admissible populations. Thus the ratio between the enrollment composition and the admissible population composition gives the propensity index. For example, if 25 percent of the total enrollment belongs to group one and only 10 percent of the total admissible population belongs to the same group, the propensity index is 250 (25/10 × 100). If instead of 25 percent only 8 percent of the enrollees belong to group one, then the index is 80. In these two cases, the 250 index indicates this group has a higher propensity by a factor of 2.5 than the average, and 80 is a below-average index. An index of 100 is considered average.

In the example depicted in Figure 3.10, the analysis has been done for one metropolitan statistical area: Cleveland, Ohio. Although it is possible to conduct a similar analysis at the state level, it is important to note that the state is not homogeneous, and the enrollment population tends to differ according to the area from which it originates. A student from Cleveland differs from one from Cincinnati or Appalachia. Therefore, to limit the analysis at the metropolitan area level makes sense. When reading this map,

Figure 3.10. Propensity to Enroll by PRIZM Group and Block Group, Cleveland

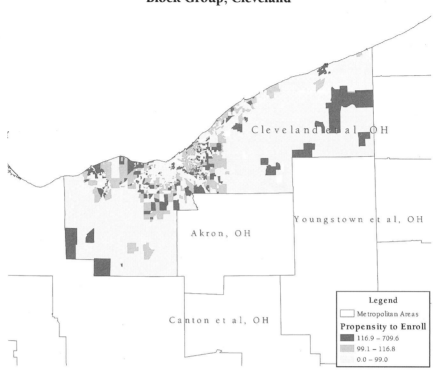

one clearly sees that students living in some areas are more inclined to enroll than others. However, something else can be analyzed. Figure 3.10 shows all the block groups in the Cleveland area classified in three propensity groups—high, medium, and low—independently of the number of enrollees in each block group. In fact, there are some block groups that do not contribute to enrollment at all. Figure 3.11 shows only those block groups where there is some enrollment.

By comparing both maps, one can identify areas with high propensity and null enrollment and consider them as potential areas for growth, or areas where recruitment efforts need to be strengthened.

Another way to evaluate the student propensity of enrollment is by analyzing the individual characteristics of the student population along with their geodemographics. In this case, each historical student record and its socioeconomic, demographic, and lifestyle characteristics are analyzed using sophisticated techniques. We have been working on creation of models to estimate the probability of enrollment of an inquiry and of an admitted student on the basis of geodemographics, academic ability, number and type

Figure 3.11. Block Groups with Enrollees, Cleveland

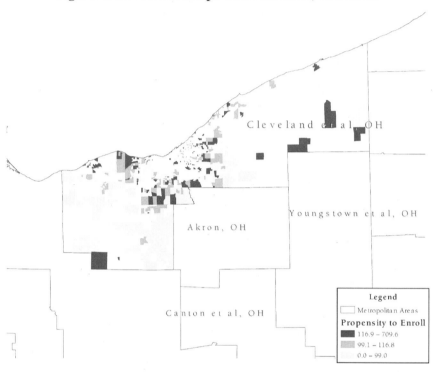

of contacts with the university, and interest in the university. These models are created using techniques such as principal components, logistic regression, decision trees, and neural networks. Though several models have been created and are already in use, we continue working to refine the geographic components of these models. So far, we know that academic ability, number and type of contacts with the university, and interest in the university are factors with predictive power. From the geodemographical point of view, distance to campus, educational attainment, type of occupation, and income-related variables are also predictive.

Another type of mapping (which is not presented, although we have done some work on it) is use of spatial analysis techniques such as spatial interpolation. This is warranted primarily when the phenomenon to map is supposed to have some spatial correlation. For example, since we know that distance from the student home address to campus is an important and limiting factor that influences the probability of enrollment, mapping this component of the probability makes sense. Nonetheless, it is also possible to use this spatial interpolation as an aid to visualize and understand other phenomena that are not necessarily spatial. For example, by using *kriging* or any other spatial interpolation technique to assess ACT scores, it is possible to create a map of this score that, when overlaid with school district boundaries, can reveal some correlation between the two. The isolines created with the ACT scores follow closely the school district boundaries.

Conclusion

This chapter presents several methods of mapping student enrollment. One should note that the data, scale, and units used in each method are suited for different purposes. The most appropriate unit to describe and understand the socioeconomic characteristics of an area is the census block group. Some examples use raw data and simple analysis, while others require more elaborate data transformation and complex analysis. These examples only scratch the surface of what can be done using GIS in examining the spatial component of enrollment data. We have come a long way, and there is more terrain to cover. The possibilities are enormous, and the future is promising.

Reference

U.S. Census Bureau. *Census 2000 Basics.* Washington, D.C.: U.S. Government Printing Office, 2002.

MANUEL GRANADOS *is a research associate in the Office of Undergraduate Admissions and First-Year Experience at the Ohio State University in Columbus. He holds master's and Ph.D. degrees in geography from Ohio State.*

*This chapter shows how institutional researchers can
make use of GIS to improve the quality of their surveys
and extract more information from their results.*

Integrating GIS into the Survey Research Process

David Roy Blough

When encountering geographic information systems (GIS) for the first time, many people begin mapping whatever data are at hand, exploring the insights that a geographic perspective adds to information. After the initial enthusiasm, though, they inevitably begin to ask how GIS can enhance the regular tasks and core responsibilities of their occupation. Unfortunately, fitting GIS into existing workflows requires more than simply becoming familiar with a specific GIS software package. This chapter demonstrates how GIS can be integrated into one of the central activities of institutional researchers: survey research.

Survey research has been a prominent part of institutional research for at least a decade, with uses ranging from assessment and enrollment management to image studies among internal and external stakeholders (Kroc and Hanson, 2001). Survey research has been among the more popular workshop topics for members of the Association for Institutional Research, or AIR (Lindquist, 1999) and is currently one of five subjects taught at the AIR Summer Institutes (Association for Institutional Research, 2003). Institutional researchers may design and implement their own surveys, use predesigned surveys on topics such as retention or freshmen satisfaction, or participate in national or multi-institutional assessments (Borden and Owens, 2001). For example, in 1991 the majority of institutions conducting research on admitted students developed a survey in-house, while roughly a fourth made use of survey instruments from the College Board or ACT (Meabon, Kelley, and Jackson, 1994).

As policy makers increasingly emphasize assessment and accountability, surveys will continue to play a role in institutional research. Yet since

NEW DIRECTIONS FOR INSTITUTIONAL RESEARCH, no. 120, Winter 2003   © Wiley Periodicals, Inc.

budgets remain strained at many colleges and universities, institutional researchers must get as much information out of surveys as they can. When integrated into the research process, GIS can increase the effectiveness of surveys.

As described elsewhere in this issue, GIS is a data management and analysis tool that can map the location and attributes of data and analyze spatial relationships among them. GIS is widely used in natural resources management, facilities planning, transportation routing and logistics, and geodemographic analysis, among other fields (Longley, Goodchild, Maguire, and Rhind, 2001). In institutional research, GIS has been used to map the locations of current students or applicants, visualize demographic change in communities, add geodemographics to enrollment models, and explore locations for new campuses (see, for example, Acker and Brown, 2001; Harrington, 2000; Pottle, 2001; Mailloux and Blough, 2000; Wu and Zhou, 2001).

Nevertheless, GIS is not often used as a tool in survey research. Unfamiliarity with GIS may be part of the reason, but one obstacle is incorporating GIS into the survey research process. GIS can make some general contributions to most survey research projects and is especially useful in surveys of external stakeholders, such as prospective students, parents, employers, or members of the wider community. Taking full advantage of the capabilities of GIS requires an understanding of the specific situations where it is most useful and its integration into the survey research process. The goal of this chapter is to assist institutional researchers in both of these aims.

A quick review of the survey research process is in order (for example, Churchill, 1988). Any research project begins with identifying a problem, phrasing a research question, and selecting a method for answering that question. If survey research is the appropriate methodology, then the first step is to define the population to be surveyed and acquire a sampling frame, or a complete list of all the subjects in the population. Then a survey is designed, a sampling method is chosen (usually simple or stratified random sampling), and a sample is drawn from the sampling frame. The survey is fielded through postal mail, over the telephone, or using e-mail and the Internet. The achieved sample (subjects who respond) is analyzed, and data findings are posted and reported.

GIS can contribute to at least three stages of the process: survey design, analysis, and reporting. In particular, it can be used to better understand the population being surveyed, make survey estimates more precise and representative, analyze regional patterns, and make maps to present findings. The next sections of this chapter illustrate each of these contributions, with examples taken from survey research projects conducted within a large public university system in Wisconsin. In some cases, examples are composites of more than one project. Since GIS does not benefit every survey research project, each section discusses the circumstances in which it is most useful.

Enabling the Survey Research Project

Making use of GIS in survey research requires geographic information about the survey subjects. For a survey of prospective students, the desired geographic information is usually the survey subject's place of residence. For a survey of current students, the permanent home address might be of interest. For a survey of area employers, the important geographic information might be the location of the company.

Typically, geographic information is available for the sampling frame in the form of a postal address, which can be geocoded as a point location in a GIS. Any survey conducted by mail has a postal address for each subject in the sampling frame. If the survey forms are coded to match records in the sampling frame, then the postal address can be linked to records in the achieved sample as well.

Surveys conducted by telephone interview, e-mail, or the Web may not have a postal address for every subject, making it more difficult to infer geographic locations. A telephone area code plus exchange can be used as a geographic identifier for land-line phone numbers, but cell phone use is increasingly rendering this method inadequate. E-mail addresses are even more problematic; the domain of an Internet service provider (ISP) conveys information about the location of the e-mail server but reveals little about the location of the e-mail user. Despite these obstacles, telephone, e-mail, and Web surveys can still collect geographic information for the achieved sample through the survey itself, by asking each subject for his or her zip code.

As long as geographic information is available for subjects in either the sampling frame or the achieved sample, GIS can contribute to the survey research process. As the next sections illustrate, some uses of GIS require geographic information in the sampling frame, some require geographic information in the achieved sample, and some require both.

Understanding the Population Being Surveyed

If the institutional researcher has access to the sampling frame of a survey, examining it with a GIS can provide a good understanding of the population being surveyed. Making use of GIS at the very beginning of the survey process can improve the design of the survey and prevent costly mistakes.

Common mishaps in survey research are to ask either the wrong questions or the wrong people, or both. It is easy to stumble into such a pitfall when the sampling frame does not adequately capture the intended population, or when the population is not well understood.

Consider an example from a four-year college in Wisconsin that serves in-state residents. The college wants to assess its image among college-bound high school students. It purchases a mailing list from an academic testing service, but the mailing list includes a substantial number of out-of-state students (from as far away as Georgia and California). The response rate from the out-of-state students—most of whom presumably have never even heard of the college—is too low to be usable. Resources are wasted surveying people who are not in the intended population.

Another example comes from a college located in a rural Wisconsin town. In developing a master's degree for health care professionals, the college surveys members of a professional organization who work in Wisconsin and Minnesota. These health care professionals are all located in major cities, and they express little interest in commuting several hours to attend classes in a small town, especially when local urban institutions already have comparable programs. The college learns about the level of interest in this group but misses an opportunity to learn about other educational preferences or behaviors.

In the first example, a miscommunication between the college and the testing service led to an incorrect sampling frame and a survey of the wrong people. In the second example, the right population was surveyed but was poorly understood. In both cases, a quick examination of the sampling frame in a GIS could have prevented these problems. The large number of out-of-state students from the first example would have been readily apparent on a map of the sampling frame, and a correct sample focusing on in-state students could have been requested. In the second example, seeing that health care professionals reside in urban areas might have influenced the design of the survey. The subjects could have been asked about their interest in distance education, or about their perceptions of competing degree programs closer to home.

Either problem could have been detected without a GIS through close examination of the sampling frames. A map, however, makes some errors (the presence of out-of-state students) immediately obvious. A map also encourages creative thinking about who is in the population and how they might respond to the survey. Examining the geography of populations during the survey design phase helps prevent costly missteps in the research process.

More Precise Survey Estimates

Surveys are conducted to gather information about a population and groups within it. If the responses to the survey are not representative of the population, then the results are not statistically reliable. If they are not precise, then the results are not useful. Two methods of addressing these issues— stratified sampling and post hoc sample balancing—are usually applied with respect to demographic characteristics such as sex or race. However, these methods can also make survey results more geographically representative.

The first method, stratified sampling, allows precise estimation of groups within the population and is carried out before the survey is fielded. Typically, the need to stratify a survey on the basis of geography is not great enough to justify the cost. On the other hand, post hoc sample balancing, or weighting the results of a survey after it is collected, is cost-effective, frequently desirable, and made easy in a GIS. Either method can be applied whenever geographic information exists for subjects in both the sampling frame and the achieved sample.

Stratified Sampling. A research project often seeks to understand the opinions of various groups within a population. If one group is relatively small, then a simple random sample may not yield enough subjects to allow a statistically precise estimate of their attitudes and preferences. For example, if Native American students make up a small fraction of the student body, a simple random sample may have too few Native Americans to yield statistically meaningful results for that group. A solution to this problem is to generate a stratified random sample by drawing a simple random sample from each group, or stratum, within the population. In this way, smaller groups (such as Native Americans) are surveyed in greater concentration, ensuring an adequate number of respondents in each group.

A sample can be stratified by geography as well as by demographic characteristics such as race. If a university surveys college-bound high school students and wants to compare state residents against out-of-state students, it may choose to stratify the sample to make sure enough out-of-state students are surveyed. However, GIS is not likely to be necessary for such a simple stratification, since in-state and out-of-state students can be distinguished easily by postal address.

GIS is more useful for stratifying on complex geographic criteria. For example, if an urban commuter school wants to understand the experiences of students who travel an hour each way compared to those of students who live nearby, mapping the students' residences might be necessary to divide the sampling frame into the two strata.

Usually there is little need to geographically stratify a sample. Stratified samples are necessary if a group within the population is very small and it is important to have a precise estimate about that group. In institutional research, groups based on geography (in-state versus out-of-state students, as an example) usually are large enough not to require stratified sampling. Moreover, stratified surveys typically have higher data collection costs since more people must be contacted to ensure that strata with a small number of subjects are adequately surveyed. Unless a survey is collected by e-mail or the Web, more subjects translate into more expense. The need for stratification by geography is unlikely to be great enough to justify the extra cost.

Post Hoc Sample Balancing. Even if all groups in a population are large enough to not require stratified sampling, the rate of subject response may vary with the group. If this occurs, and if attitudes and preferences differ among groups, then the overall results of the survey may not represent the population as a whole. Consider a survey of a population that is 50 percent male but with a higher response rate among females. If the achieved sample is 60 percent female, and some opinions tend to differ by sex, then the opinions of females are overrepresented in the survey.

Response rate can also vary with the geographic region. In institutional research, surveys of external stakeholders are especially susceptible to this problem because awareness and perception of a college or university can vary considerably with the residential location of the respondent.

A common, cost-effective way to alleviate variation in response rate is post hoc sample balancing, also referred to as poststratification. Poststratification assigns a weight to each respondent on the basis of characteristics that are known for both the population (or in practice, the sampling frame) and the achieved sample. Once the weights are applied, the characteristics of the achieved sample match those of the overall population.

In general, the total sample weight for each subject takes the form TSW $= P_c/R_c$, where P_c is the proportion of people in the population who have the same combination of characteristics as the subject, and R_c is the proportion of people in the achieved sample with those same characteristics (Korn and Graubard, 1999). The total sample weight increases or decreases the influence of each respondent, to bring it closer to the relative proportion of individuals with similar characteristics in the population. If any proportions are the same in the population and the achieved sample, the total sample weight equals 1.

For example, consider a survey of college students in which sex and enrollment status (full-time or part-time) are known for each subject in the sampling frame. As the survey is collected, women and full-time students respond at higher rates than men or part-time students. Full-time women students are only 20 percent of the population, but they make up 30 percent of the achieved sample; for part-time male students, the proportions are reversed. For the other two combinations, full-time males and part-time females, the proportions are 15 percent/20 percent and 35 percent/30 percent respectively.

Using post hoc sample balancing, full-time female students (the most overrepresented group) are each assigned a total sample weight of 0.20/0.30 = 0.67. Part-time males (the most underrepresented group) each would have a weight of 0.30/0.20 = 1.5. The remaining groups would have weights of 0.15/0.20 = 0.75 and 0.35/0.30 = 1.17. Thus a part-time male student is counted as one and a half respondents, while a full-time female student is counted as two-thirds of a respondent. In this way, the proportions by sex and enrollment status in the achieved sample are adjusted to match the overall population.

Just as the achieved sample can be weighted by gender or enrollment status, it also can be weighted by geographic region, ensuring that survey results are geographically representative of the population. Having a survey that is geographically representative can be important in a study assessing the awareness of a place or institution.

Consider the experience of a college campus in western Wisconsin that enrolls a large number of students from neighboring Minnesota. This college surveys prospective students from the region; the overall results show both high awareness and a positive perception of the college. Closer examination, though, reveals that in-state response to the survey is much higher than response from subjects in Minnesota. It turns out that awareness among Minnesota respondents is low, and perception of the college is less positive as well.

Should the campus ignore the geographic difference in response rate and conclude that it enjoys a high degree of positive recognition in the region? Doing so would be misleading. The overall awareness is overrepresentative of in-state prospects. Yet Minnesotans are an important part of the population the campus wants to understand.

Should the campus report awareness and perception of Minnesota prospects separately from Wisconsin prospects? Certainly, but an estimate of overall awareness may be desirable, too.

Should the campus expend additional time and money to resurvey the out-of-state prospects? This is the best option if resources are available and the same group of prospects can still be reached. However, meeting either of these criteria can be difficult. Money may not be available for another survey, and enough time may have passed that the prospective students have already enrolled at a college or university. In addition, repeatedly contacting subjects who prefer not to answer a survey could be construed as harassment.

Post hoc sample balancing is a practical way to address all of these issues. Weighting the achieved sample by geographic region allows calculation of an overall estimate of awareness that represents the relative influence of in-state and out-of-state subjects in the population. In addition, estimates for each group (in-state or out-of-state) can still be calculated. Finally, poststratification does not involve the resources, timing issues, and ethical dilemmas of resurveying nonrespondents. GIS is especially useful for weighting surveys by geographic region because it facilitates delineating regions on the basis of geographic features other than state lines. Once the sampling frame is mapped by postal address or other geographic information, it becomes a simple matter to delineate regions that correspond to the distribution of the population.

In some research projects, institutional researchers do not have direct access to the sampling frame; they might be participating in a survey that is administered through a third party. Post hoc sample balancing can still be performed, though, if the owner of the sampling frame is willing to furnish summary statistics about the frame. For example, if the list owner gives a count of the number of males and females in each zip code, and the survey collects sex and zip code from each respondent, then the achieved sample can be balanced against the sampling frame by sex and geographic region.

Analyzing Regional Differences and Spatial Behavior

The uses of GIS discussed thus far have relied at least in part on geographic information about the sampling frame. However, once a survey has been designed, fielded, and poststratified (if necessary), analysis with GIS can rely solely on geographic information in the achieved sample. In other words, geographic analysis can be conducted on a survey even in a situation where the researcher has no access to the sampling frame, so long as geographic information such as zip code has been collected for the achieved sample through the survey instrument.

GIS is particularly useful in two types of analysis: statistical differences among geographic regions, and estimating potential market areas. Statistical differences among regions are likely to arise in examining awareness and perceptions among external stakeholders. Estimating potential market areas, which takes advantage of the full analytic functionality of GIS, is useful for anticipating demand for new programs tailored to nontraditional student populations.

Statistical Differences Among Regions. As was mentioned previously, survey research often seeks to uncover differences in attitude or perception among respondent groups. These groups can be based on demographic characteristics; for example, when choosing a college, do males say "campus life" is more important than females? Or, groups can be based on an attitude or preference: Do those students who say they are not satisfied with course scheduling options work more hours a week then students who are satisfied?

A third basis for a group is location: Do prospective students who live nearby have different opinions than those living farther away? With this type of question, GIS is a useful tool for grouping respondents into geographic regions. An example illustrates a situation where regional differences offered insights that otherwise would have been missed.

A college admissions office was concerned about "canceled" students—new freshmen who initially agreed to enroll but later decided not to attend. The college had several hypotheses about why students canceled, among them the suspicion that students did not want to live in the town where the campus is located.

In a telephone survey, canceled students were read a series of statements describing the college, its faculty, the admissions process, and the town in which it was located. Then each student was asked to rate, on a five-point scale, how important each statement was in his or her decision about which university to attend. Later in the interview, students were asked to rate how well each statement matched their perception of the college. For analysis, statements addressing similar issues were combined into factors, and a repeated measures test (Kirk, 1995) was used to detect significant differences between the importance and perception of each factor.

From the results of the tests, factors were characterized in one of three ways. Factors with an average importance rating that were greater than their perception rating were considered "image opportunities" because the college was perceived as not possessing these attributes to the degree that they were important to respondents. Factors with an importance rating less than the perception rating were classified "image strengths" since in these cases the college was perceived as exceeding the level of importance assigned to the factor. Finally, for some factors, importance was not statistically different from perception. Here, perception of the college matched the level of importance for those attributes—an "image match."

To examine the college's hypothesis about the desirability of its location, the College Town factor included statements such as "The campus is located in a scenic city," "Restaurants and stores are located within walking distance of campus," and "The city is located in a community that has interesting things for college students to see and do." When the College Town factor was analyzed over all respondents, it was found to be an image match. The attributes of a College Town—being in a scenic city and so on—characterized the college to the same degree that they were important to canceled students. This finding was considered good news for the college because it suggested that shortcomings of the campus location were not meaningful reasons for these students to cancel their admission.

Even so, the question remained whether respondents throughout the state shared this image. Using GIS, the respondents were grouped into eleven regions centered on major metropolitan areas and having approximately equal numbers of respondents, enough for statistical testing. For convenience, regions followed county boundaries, excluding counties that had no respondents.

Results of the repeated measures test for each region revealed geographic variations in the image of the College Town factor (Figure 4.1). For respondents from most regions, including the region where the college was located, the College Town factor was still an image match. In a few regions in the central and northeastern parts of the state, the factor was an image strength. In southwestern Wisconsin, however, the College Town factor was rated an image opportunity; respondents in that region felt college town amenities were important but perceived a lack of them at the college's location.

The benefits of adding geography to this analysis were twofold. First, looking at responses within regions uncovered information that otherwise would have been missed. The average image of the college town was satisfactory, but this average obscured important differences among respondents. Second, the regional picture offered a greater understanding of how the college was perceived in different parts of the state. Students from southwestern Wisconsin may cancel their admission for different reasons than what applied to students from other regions. This understanding of the image of the college in a region could shape the communication strategy with potential students from that area.

Figure 4.1. Importance and Perception of the "College Town" Factor Among Canceled Students

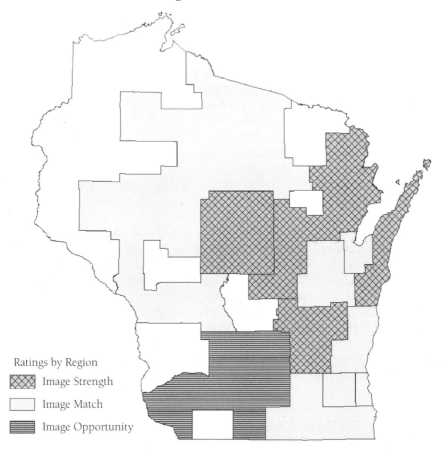

Ratings by Region

Image Strength

Image Match

Image Opportunity

Location has an inherent influence on several attitudes and preferences. First, the awareness of a place or an institution tends to be higher in the immediate vicinity of an institution and lower among people farther away. An example of this situation is the case discussed earlier where a Wisconsin college had lower awareness among prospective students in Minnesota than among in-state prospects. Second, perception may differ from region to region even if awareness is high in all locations, as in the example of the image of the college town just described. Third, the spatial relationship between the respondent and the place or institution may differ from region to region. As intuition would suggest, high school students who say a college is "too close to home" tend to live closer to the college than those who say the college is "too far away"—although there are plenty of individual exceptions.

These attributes of awareness, perception, and spatial relationship are likely to vary by geographic region among prospective students or others outside the college or university. Applying GIS to survey analysis can help uncover these regional differences.

Estimating Potential Market Areas. The contributions of GIS to survey research discussed thus far have not involved much higher-level GIS analysis. Mapping a mailing list or delineating regions may be quicker and easier, but a GIS is not the only way to get the job done. Where survey research can take full advantage of GIS functionality is in analysis of market areas.

In higher education, market areas (often called service areas) are the regions from which a college draws most of its students. For most colleges, the market area for traditional undergraduate programs is well understood; typically, a college draws heavily from the same set of high schools every year, with minor variations. Likewise, a college with established programs serving returning adults or other nontraditional students also may have a clear sense of where their students are coming from. A college or university may find it difficult, however, to anticipate the size and location of the potential market for new programs or for existing programs being extended to nontraditional students. In such a situation, survey research and geographic analysis can be combined to gain insight into the potential market, as another example illustrates.

A consortium of universities in the upper Midwest plans to offer a specialized graduate degree program for working professionals. Most of the coursework for the degree will be delivered through online distance education, but some face-to-face instruction is an essential part of the program. One concern is where to hold the face-to-face classes; should they be hosted at the participating institutions, or would alternative locations be more convenient?

The consortium surveys members of a professional organization who are living and working in the upper Midwest. The survey describes the potential program and asks for overall level of interest and preferences for content, scheduling, and delivery. The survey asks for some basic demographics, including the respondent's zip code. Finally, to ascertain commuting preferences, the survey asks how far the respondent would be willing to drive to attend face-to-face classes in the proposed program.

The commuting question, when analyzed along with the respondent's zip code, yields information about the potential market areas for the new degree program. Using GIS, the location of each respondent is estimated using the approximate center (or centroid) of the respondent's zip code of residence. Then a commuting area is defined around each respondent's residence on the basis of how far the respondent is willing to drive to attend classes in the program. (To account for differences between rural and urban areas, driving distance is measured through the road network.) Repeating this process creates a unique commuting area for each respondent according

Figure 4.2. Estimated Commuting Areas for Survey Respondents

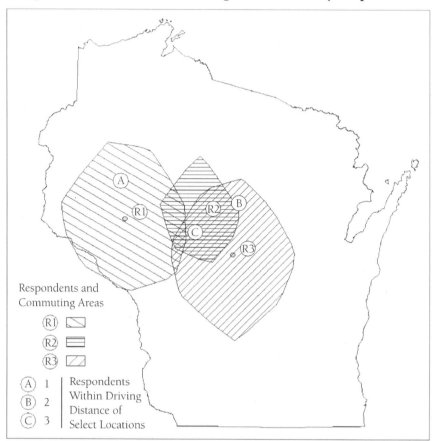

Respondents and
Commuting Areas

- (R1) ⬚
- (R2) ▤
- (R3) ▨

(A)	1	Respondents
(B)	2	Within Driving
		Distance of
(C)	3	Select Locations

to his or her location and stated commuting preference. For example, in Figure 4.2, respondent R1, from Eau Claire, Wisconsin, indicates a willingness to drive sixty miles to attend monthly face-to-face classes in the proposed program. This respondent's potential commuting area extends over much of western Wisconsin. Respondent R2, from Unity, Wisconsin, is willing to drive only forty miles, while respondent R3, from Wisconsin Rapids, is willing to drive sixty miles.

Once the commuting area for each respondent has been defined, the commuting areas are converted into grid (raster) form, overlaid, and summed together. The resulting grid reflects the number of respondents who would be willing to attend classes at each location. Put more precisely, the value of each grid cell reflects the number of respondents who have that location in their commuting area. In Figure 4.2, a grid cell at point A has a value of one because it is within the commuting area of only respondent

**Figure 4.3. Convenient Locations for Attending Monthly
Face-to-Face Meetings**

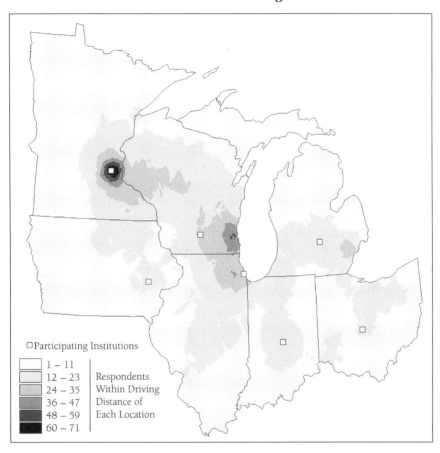

R1. A grid cell at point B equals two, since it is within driving distance
of respondents R2 and R3. Point C is in the commuting areas of all three
respondents.

Applying this method to all respondents generates a map showing the
locations that are convenient to the largest number of people (Figure 4.3).
The Minnesota institution, for example, is located in an area that is within
driving distance of many respondents, while the Iowa institution is not.
Furthermore, the Detroit area in southeastern Michigan is convenient to
more respondents than the area around the Michigan institution.

In part, locations that are convenient to the most respondents reflect the
underlying geographic distribution of the surveyed population. Cities with
many respondents (Minneapolis–St. Paul, Milwaukee, Chicago, Detroit) are
convenient to more people. The transportation network also affects which

locations are more convenient. A large part of western Wisconsin is convenient to many respondents because of an interstate highway that runs east and west through the region. Finally, the individual commuting preferences of respondents are a factor. Respondents in urban areas are less willing to drive a long distance than respondents in rural areas are, perhaps from either dislike of urban traffic or greater expectations of having a convenient location close by. Taken together, these three factors combine to present a picture of the potential market areas for the proposed degree program.

Studying potential market areas with a combination of survey research and GIS entails several benefits. First, the combined method tells us more about the spatial pattern of demand for the program than the survey alone. The survey question "How far would you drive. . . . ?" tells us only that on average respondents would be willing to drive fifty miles to attend classes. This statistic is helpful, but it reveals little about who would travel to the institutions offering the new degree program, or where respondents would prefer classes be held. Combining this survey question with each respondent's location and applying analytical tools from a GIS yields insight into locations that are more accessible or appealing.

Communicating Findings

Integrating GIS into survey research has many behind-the-scenes benefits: visualization of the population being surveyed, extension of methods that enhance precision and representation of survey estimates (such as stratification and weighting), statistical tests of regional differences, and analysis of potential market areas. The most visible benefit of GIS, however, is its ability to make maps.

Despite the appeal of maps, caution must be taken in using them. Cartographic literacy is not always at the same level as graphic literacy, whether for the map reader or the map creator. Cartographers have well-established methods for conveying information in ways suited to human visual perception (see, for example, Dent, 1999). Even with good cartography, though, maps can be confusing to an audience more accustomed to interpreting charts, graphs, and tables.

Maps are most effective when their use is guided by a few common-sense principles. The most obvious one is to make use of a map only if it illustrates a point with significance to the overall study. As a corollary to this principle, the point being illustrated should involve geographic location or variation; otherwise, a chart, graph, table, or text is more appropriate. Finally, it is safest to make a map that conveys just one message or idea. People who are used to comparing two variables on a chart or graph often have trouble drawing similar comparisons on a map. Even when a comparison can be displayed on a single map, following well-established cartographic conventions, two or more separate maps are often more easily understood by an audience.

Conclusion

GIS is more than making maps, just as survey research is more than asking questions. If GIS is planned into a survey project from the beginning, it can enhance the research process before the first map of results is ever made. Survey research is a core activity of the institutional researcher and is likely to remain so. As this chapter has shown, institutional researchers can make use of GIS to improve the quality of their surveys and extract more information from their results.

References

Acker, R. J., and Brown, P. H. "The Use of Geographic Information Systems (GIS) in Institutional Research." Presented at the 41st AIR Forum, Long Beach, Calif., June 3–6, 2001.

Association for Institutional Research. "Foundations for the Practice of Institutional Research Institute." [http://www.airweb.org/page.asp?page=406]. Accessed Mar. 28, 2003.

Borden, V.M.H., and Owens, J.L.Z. *Measuring Quality: Choosing Among Surveys and Other Assessments of College Quality.* Washington, D.C.: American Council on Education and Association for Institutional Research, 2001.

Churchill, G. A., Jr. "Basic Marketing Research." (Dryden Press Series in Marketing.) Chicago: Dryden Press, 1988.

Dent, B. *Cartography: Thematic Map Design* (5th ed.). Boston: McGraw-Hill, 1999.

Harrington, R. A. "Geographical Characterization of Applicants Using the Admitted Student Questionnaire." Presented at the 40th AIR Forum, Cincinnati, Ohio, May 21–24, 2000.

Kirk, R. E. *Experimental Design: Procedures for the Behavioral Sciences* (3rd ed.). Pacific Grove, Calif.: Brooks/Cole, 1995.

Korn, E. L., and Graubard, B. I. *Analysis of Health Surveys.* New York: Wiley-Interscience, 1999.

Kroc, R., and Hanson, G. "Enrollment Management and Student Affairs." In R. D. Howard (ed.), *Institutional Research: Decision Support in Higher Education.* (Resources for Institutional Research, 13.) Tallahassee, Fla.: Association for Institutional Research, 2001.

Lindquist, S. B. "A Profile of Institutional Researchers from AIR National Membership Surveys." In J. F. Volkwein (ed.), *What Is Institutional Research All About?* New Directions for Institutional Research, no. 104. San Francisco: Jossey-Bass, 1999.

Longley, P. A., Goodchild, M. F., Maguire, D. J., and Rhind, D. W. *Geographic Information Systems and Science.* New York: Wiley, 2001.

Mailloux, M. R., and Blough, D.R. "The Who and Where of Adults in Higher Education: Merging Modeling Techniques with GIS to Understand Participation in Credential Courses." Presented at the 40th AIR Forum, Cincinnati, Ohio, May 21–24, 2000.

Meabon, D. L., Kelley, J. W., and Jackson, D. L. "Admitted Student Market Research: A National Perspective." *Journal of Marketing for Higher Education,* 1994, 5(2), 17–29.

Pottle, L. "Geographic Information System (GIS) Applications at a Multi-Site Community College." Presented at the 41st AIR Forum, Long Beach, Calif., June 3–6, 2001.

Wu, J., and Zhou, Y. "Building a University Student Enrollment Database with ArcView GIS: A Step-by-Step Demonstration." Presented at the 41st AIR Forum, Long Beach, Calif., June 3–6, 2001.

DAVID ROY BLOUGH *is an institutional planner for the University of Wisconsin System Administration and has a Ph.D. in geography from the University of Wisconsin–Madison.*

Growth and technology have driven a public university to build a GIS database for its facilities and space management. This chapter reviews the benefits and challenges that come with this implementation.

Building a GIS Database for Space and Facilities Management

Nicolas A. Valcik, Patricia Huesca-Dorantes

The use of geographic information systems (GIS) in the public sector has increased over the last couple of decades. New applications of the software vary in sophistication, as was mentioned by Martin (1991), from satellite imagery to a subgeographic operation based on computer-aided design (CAD) systems. Creation of digital maps and data offers "highly sophisticated manipulation options" (Martin, 1991, p. 13). Historically, GIS has been used to describe "isolated applications that enabled urban planners and policymakers to visualize certain data-sets on a map" (van de Donk and Taylor, 2000, p. 128). More recently, GIS has been used in a variety of fields, from crime prevention, health care, and urban renewal (Martin, 1991) to environmental policy and traffic control (Meijer, 2002). Higher education, through the division of institutional research, became interested and consequently involved in this trend. This paper illustrates how the University of Texas at Dallas (UTD) took the challenge and implemented a GIS to fulfill the demands on space and inventory management.

Challenges

Like any other public university, UTD is subject to mandated reports. Accurate measurement and reporting of space is important to a public university because of the direct relationship of accuracy to funding received from the federal and state governments for projects such as new buildings or renovations.

NEW DIRECTIONS FOR INSTITUTIONAL RESEARCH, no. 120, Winter 2003   © Wiley Periodicals, Inc.

Many state institutions currently use the *space projection model*, approved in 1992 by the Texas Higher Education Coordinating Board (THECB), which "predicts the net assignable squared feet of educational and general space and institution needs in five categories: teaching, library, research, office and support space" (THECB, 2000, p.1). The coordinating board uses this model "as part of its review process in the consideration of facility projects that would generate new space" (THECB, 2000, p.1) and as part of the funding formulas for the general academic institution, the assistance fund allocation formula, and infrastructure formula.

Federal indirect cost reports are tied to space and inventory management since funds for research, in the form of contracts and grants, are approved on the basis of those reports. It works as follows. Space usage surveys are updated through paper questionnaires. The survey is conducted once every three years, mailed out to departments and completed by departmental administrative assistants or program heads.

This process generated a certain amount of stress and usually ended with the academic deans being contacted by one of the executive directors in the central administration. Eventually, once the paper surveys were returned to the Office of the Controller, their personnel entered all the data into the old database system.

In many institutions, calculations of square footage have traditionally been accomplished by hand and then input into a database system. The process usually takes employees several days to measure rooms every time a building is constructed or renovated.

Increasing Efficiency and Accuracy

Before any type of spatial database was constructed, the information was not kept electronically. On the one hand, keeping a hard copy of records was done with minimal training, and usually hard copies "have an important impact on availability of data for accountability" (Meijer, 2002, p. 41). On the other hand, this method consumed time, paper, and physical space (in the form of files). Later in the year, the data entered were audited by the Office of Strategic Planning and Analysis (OSPA). By that time, errors were discovered and the data accuracy was lagging behind the rapid development of the university. Quite often, OSPA had to synchronize the facilities database with the other university database systems.

The Need for GIS

The Office of the Controller, when prompted by the deadline for the National Science Foundation's *2001 Indirect Cost Survey,* placed a formal request for assistance with OSPA to reload the data into the old facilities database system. This request was rejected because loading new data into an old system would not solve the inherent problems. This issue—along with the enrollment growth that UTD was experiencing—triggered the need to design a physical inventory and management system larger than the one currently in existence. The result was the Spatial Inventory Database, or SID (2002 provisional patent filed). Among other features, this system includes a georeferenced database linked to nonspatial data through the use of geographic information systems.

The Design of the Spatial Inventory Database

In SID, queries and reports can be produced more quickly on a personal computer application, and the process does not require the services of a programmer to manufacture a given set of data. Before undertaking the task of building a spatial inventory database, an organization should conduct a needs assessment. These are some of the questions that can help when planning for the database:

- How many facilities does the institution have?
- What type of database platform can process the number of records needed?
- How does the user carry out data entry and report the data?
- Which departments are involved in the implementation plan, and to what degree?
- Is the technology in place to accomplish the strategy and goals?
- Are the personnel trained to use the technology?
- How much training is needed to run and maintain the database?

If any of these questions are not addressed, the project might run into problems on design, or worse on implementation.

OSPA is acutely aware of the difficulties other departments experience with data entry and manipulation. Its personnel possess programming capabilities and can create reports directly from the mainframe to fit their own specifications. They are one of the largest end users of virtually all the databases on the mainframe and on personal computers platforms at the institution. Despite their programming skills, many members of OSPA have found, from their own experience and through conversations with other departments, that using the databases is time-consuming and rather difficult.

Strategic Planning, along with the Controller's Office, began to build the initial SID prototype once the proposal for a new design was accepted. Beyond the technical considerations of constructing a computer database, issues regarding data use, security, departmental responsibility, and administrative expectations had to be met prior to commencement of the project. In an academic setting, political issues have the potential to be just as debilitating if not handled appropriately.

Reasons to Use GIS

The use of GIS is broad. For instance, it accelerates the process of reporting and at the same time creates more accurate reports. There was great demand for integration of GIS for campus mapping along with facility inventory control. SID integrates the ESRI ArcSDE extension and ESRI ArcView 8.1 and allows GIS analysts to manage spatial data in a database management system. That is, it can convert drawings with dimensions into Microsoft SQL Server 2000 data.

Accurate Space Measurements. All facilities, floor by floor, are mapped and stored in the form of shape files, in the ESRI ArcSDE server. These shape files contain, among other information, accurate perimeters, areas, and volumes, in a system of coordinates that can be drawn from the software. The use of these coordinates gives the spatial files an accuracy that regular nonspatial files cannot deliver, especially when obtaining areas or volumes in odd-shaped rooms and capturing not only assignable but also nonassignable space, such as telecommunication closets, mechanical rooms, and infrastructure. This feature reduces the level of effort to accomplish the measurement of space as well as the updating of the files. GIS also allows three-dimensional measurements, helpful for calculating heating and cooling utility costs, with a formula based on the amount of square footage. Fire codes are dictated by square footage as well, and with GIS it is feasible to obtain accurate ratios for room capacities.

Accurate Mapping Technology. From a less critical standpoint, GIS creates accurate maps for departmental and student use. Large and small campuses can benefit greatly from the use of accurate maps. The advantage varies, from directing a supplier to a building to directing a potential job candidate to an interview. In-house employees can also benefit from the maps. Engineers and telecommunications personnel can locate the best pathway to lay new cable or infrastructure hardware for upgrades to existing systems, or for a new facility under construction. Again, using GIS can make campus planning more efficient. GIS allows forecasting of new buildings according to space needs, enrollment, campus land use, and availability. As a consequence, traffic patterns, utility layouts (electrical, water, sewer, telecommunication, gas lines), and even the cost and look of new additions can be represented with this software.

Emergency and Hazardous Control. Cut-off water valves for sprinkler systems and other fire fighting equipment can be mapped for emergency system personnel. This enables them as well as local municipal or state emergency personnel to know the location of such devices. Hazardous materials, chemicals, biological specimens, and radioactive elements can be assigned properties in the database that contain valid room numbers within the facilities. This information is input into the database for different areas affected at the university, allowing police and safety crews to access the information quickly and accurately. This record gives the police or emergency crews an inventory tied to a location, and again it allows those users to access the data by way of wireless personal digital assistant (PDA) systems. Thus risk management can know not only the kinds of hazardous materials but also which department owns the material and how and where the materials are stored. As an added security measure, camera locations, call boxes, and crime statistics can all be kept in synchrony with GIS and the database.

Easy Accessibility and Use. With the interconnectivity of the GIS to a facilities inventory database, personnel and equipment inventory can be georeferenced to the floor plans. Personnel can be assigned to multiple rooms such as faculty offices or laboratories; conversely, multiple personnel can be assigned to a single room such as an office suite. Departmental ownership can also be illustrated by feeding the information from the database to the GIS. This provides information reporting capability to security and emergency personnel, who may need to know who is actually in the building. It also allows university administrators to determine who controls space by a visual reference to floor plans that are color coded by departmental ownership.

Error Control. GIS offers error control when inputting data into the system. For instance, when scheduling, if a classroom does not exist in the geodatabase then that particular room cannot be selected. Prior to GIS implementation, scheduling was done for rooms that were not coded as classrooms or simply did not exist. The same principle applies to physical plant crews, telecommunications, and other areas that need to access information about infrastructure, ownership or personnel, and their room assignments. If physical plant workers are digging a trench for an electrical line, they can first check the Website and the GIS renderings using a PDA to see if any other potentially hazardous line (telecommunication, water, or sewer) exists in the area.

Combining Visual Basic scripting with Microsoft Access forms or Microsoft InterDev ensures greater data accuracy and simplifies data entry. For example, the Room Survey Webpage on SID adds Visual Basic scripting to simplify the process for the user and to ensure correct data entry. The Room Survey Webpage allows users to select the correct building and room number before they attempt to enter information for that area. For instance, when filling in CIP codes and percentages in a table, Visual Basic scripting ensures that the percentage totals 100; the user cannot continue to the next section if the percentage does not equal 100, or if other erroneous data are entered (such as typing a character instead of a numerical value). This forces the user to correct errors before continuing to the next screen.

How GIS Works

Once institutional needs were assessed, the database design and table layout were carefully planned. The table structure of the database is critically important for deciding how many tables are needed and what information

Figure 5.1. Organization of Room and Building Tables

ROOM TABLE		BUILDING TABLE
Building	Room No	Building
MP	2.408	MP

should be contained in each table. For the purposes of creating flexible and intricate queries, several smaller tables are preferable with specific, well-defined characteristics. The more variables that are placed into one table, the less flexible the table becomes in answering administrative questions regarding facility planning, processing information, and Webpage interface design. Relationships among tables are easily created yielding many permutations, which in turn can produce sophisticated and detailed reports tailored to the needs of the end user.

Such relations are not possible without standardization of the data. For example, the building codes in a building table must match the building codes in a room table. Between two tables, there must exist one data field that is common to both, or the tables cannot relate properly. Furthermore, the data contained within the common field or index must be consistently input or the relation will produce erroneous data. As Walford (2002, p. 14) describes it, "a more useful kind of index is one which provides a unique code for each individual occurrence of a particular kind of feature." Note the entry examples in Figure 5.1 as an illustration of this concept.

In the building table seen in Figure 5.1, the code for the hypothetical building is "MP," while in the room table the first field contains the building code and the second field contains the room number. Since the building codes in the two tables are identical, these two tables can be related for querying. Also note that room listings and master building information have been placed in two separate tables, not combined into a large master table. This is an example of how creating smaller tables yields a flexible, customized report. The linkage of tables is performed through programming code on the data reporting side of the system. Once they are linked, it is necessary to plan how users will interact with this database.

Webpages were created for specific departmental needs with access restrictions to different areas in the database, from full access to all data entry and reports to highly restricted access (that is, limited data entry and reporting capability). Webpages were designed to relate features compiled through coding in the database (for data entry control) for those users who have read-only access. The coding enables various levels of access into the system. Such a feature grants greater control over the database and can reduce the likelihood of accidental errors.

Once the pages are built, queries and reports were created in InterDev and linked to simple command buttons on the forms or through the Web to connect to the database. A query filters data to produce a table, while a report presents the information in a professional format, which can include headers, page numbers, and the day and time the report was created. Reports can be set up to download their results into a Microsoft Excel spreadsheet or in text format so that the end user can manipulate the data in any desired form. Maintaining the data in electronic format, that is, downloaded directly from the system to the desktop, is more efficient than producing hard copy. It is also more secure than sending information through e-mail or floppy disks. In SID, secured data entry is accomplished by using a Secure Socket Layer for the transmission through the Webpage to the SQL Server 2000 database.

Conclusion

Soon after the SID was put into production, users began to ask questions and make observations regarding the program's functionality. Those with limited formal computer training preferred the user-friendly point-and-click graphic user interface found on Websites to the more complex database forms. Senior administrators wanted easy access to facilities and reports that were compiled in a professional format. The property administrator wanted to use one form that would input data, produce reports, and perform editing functions on room and building tables. Departments that maintained their own databases of room and building information wanted to match their data to the larger campus database.

What started as a simple request for assistance became a large-scale endeavor to improve the university's entire space inventory structure. The development of a new space inventory system can foster a number of challenges, especially technical difficulties. Importing existing data into the new tables can create serious problems if the information is exported from the old system incorrectly. Something as mundane as a specifying the wrong character length in a column can disrupt the process. Digitizing, which is creating digital maps from the CAD or from other sources, is time-consuming and error prone (Martin, 1991). This fact creates the need for data verification procedures. Therefore, after data conversion time must be set aside to ensure the dataset is completely free of errors.

Creating SID demanded cooperation among many departments and a pooling of personnel and financial resources. Persuading these personnel that the systems have to be redesigned for the improvement of the university is an ongoing process. Designing a database system in house fueled much skepticism. Personnel who traditionally used only mainframe technology were often distrustful and threatened by the new technology of the project.

To avoid miscommunication on a project of this dimension, it is necessary to schedule meetings and maintain dialogue among all parties involved to delineate responsibilities. Communication is also critical in ensuring that everyone understands and agrees upon the purpose and implications of the project as well as the capabilities and the limitations of the new system. Those attending the initial meeting had computer skills that ranged from minimal to advanced. Some who attended had low expectations from the system and only wished to see one or two reports as a result of the effort, while others wanted a system that could maintain the data automatically and accommodate more complex queries.

Establishing the capabilities of the system before the system is completed is highly important. OSPA and the Office of the Controller kept these issues in mind when conducting a joint presentation to the senior vice president of business affairs and various other directors prior to project initialization. Careful planning and open communication channels with all parties resulted in a project that is successful for the most part, free from serious political infighting and the disappointment of flawed expectations.

The need was imperative. The potential investment returns on time and funds are very encouraging, and the impetus to succeed is sufficient to take up the challenge. Careful planning, close cooperative effort, strong lines of communication, and insightful design enabled the SID project to be used, licensed, and sold to other entities. Early results already suggest that the workload for personnel in the Office of the Controller has substantially decreased in regard to completion of the NSF indirect cost surveys and for data entry reports to other departments. The time saved can now be applied toward other endeavors. The high level of exhaustion and frustration engendered by the old space management system has been reduced through the use of SID and has also ensured greater data accuracy. Furthermore, it has yielded an unprecedented level of cooperation among departments, maximizing the effectiveness of everyone's efforts and minimizing wasteful activity and miscommunication. Above all, SID has empowered personnel to do their jobs more quickly and efficiently, which ultimately is the primary purpose of all computer systems.

References

Martin, D. *Geographic Information Systems and Their Socioeconomic Implications.* London: Routledge, 1991.

Meijer, A. "Geographical Information Systems and Public Accountability." *Information Policy,* 2002, 7, 39–47.

Texas Higher Education Coordinating Board (THECB). *Space Projection Model for Public Universities, Technical Colleges and the Lamar State Colleges.* Sept. 2000. Austin, Tex.

van de Donk, W.B.H.J., and Taylor, J. "Geographic Information Systems (GIS) in Public Administration: An Introduction to a Series of Articles." *Information Infrastructure and Policy,* 2000, 6, 127–129.

Walford, N. *Geographic Data, Characteristics and Sources.* West Sussex, England: Wiley, 2002.

NICOLAS A. VALCIK is an assistant director for the Office of Strategic Planning and Analysis at the University of Texas at Dallas. He has designed multiple databases for interfacing with end users.

PATRICIA HUESCA-DORANTES is an institutional research associate for the Office of Strategic Planning and Analysis at the University of Texas at Dallas. She has provided extensive GIS analysis on demographics.

This chapter is based on the author's experience working with geographic information systems (GIS) at the University of Arizona, along with insights gained through literature reviews and dialogue with staff at other universities.

Developing Enterprise GIS for University Administration: Organizational and Strategic Considerations

B. Grant McCormick

The focus of this chapter is on the use of GIS for "administrative purposes," which is to say, any application related to planning, operating, or managing a university campus.

A presumption is that key benefits of GIS are to be realized with a system that permeates the enterprise, links divisions, integrates data sources to create new understanding, and creates efficiencies by overcoming territorial boundaries. With this in mind, the term *GIS* should be seen not solely as the use of GIS software, but rather as a technical framework for integrating disparate datasets, bridging software formats, and responding to a plethora of administrative needs and goals. GIS functions, then, as a technological glue joining a range of software packages, custom applications and interfaces, datasets, and so on into something that might be more appropriately termed a *spatial data management system*.

The intent is not to generate an exhaustive checklist of activities for implementation; rather, it is to convey some key organizational issues that are likely to bear on most campus administrative GIS efforts.

The first section presents several unique aspects of the campus context that appear relevant to GIS initiatives, while the second section is a summary of organizational dynamics affecting GIS implementation. The third section reviews basic forms of campus GIS, and the final section presents key strategic issues for enterprise GIS.

The University Campus as a Context for Administrative GIS Use

The university campus, as a context for GIS applications, presents unique challenges and opportunities when compared with other government enterprises or the private sector. Here is a review of several institutional characteristics of relevance to how GIS may fit within an institution.

Multiple Communities. Many communities overlap to form a university setting, ranging from communities of faculty, staff, and students to neighbors, alumni, visitors, consultants, local government, and an array of special-interest communities. This translates into a challenging number of concerns, thinking styles, decision-making structures, and applications of technology. As a result, it may be difficult for a GIS initiative to effectively address all potential applications and users. Conversely, these communities represent a wealth of untapped GIS applications.

Identity, Mission, and Tradition. Universities typically value their unique identity, which is often expressed through a mission statement, celebrating treasured aspects of the physical campus, preserving history and traditions, or highlighting areas of academic excellence. Guiding notions about an institution may give insight as to possible applications, levels of support, and major hurdles. For example, a tradition of decentralization as opposed to one of central authority may influence the direction a GIS takes. The presence of a campus arboretum and a heritage of tree stewardship may suggest vegetation mapping as a potential GIS application.

Smaller Physical Scale. GIS applications have typically occurred within geographies that are relatively large when compared with the campus geography. The error in a regional-scale database may be greater than the entire extent of a university campus (that is, an error tolerance of perhaps a quarter mile in a statewide vegetation map). Maps are essential in most traditional GIS contexts because it is not physically possible, because of size or remoteness, to directly inspect the entire environment. Many standard GIS analyses (for example, determining a buffer, finding what is inside something else, and so on) don't have the same analytical impact when one has the option of directly examining the environment. Experience suggests that the ability to directly inspect the campus reduces somewhat the perceived need for map maintenance. If maps are prepared, however, the expectation of spatial accuracy appears to increase in the context of the more detailed campus environment.

Urban Core Setting. GIS has traditionally been applied in such settings as natural resource management and regional utility or land records. Municipal GIS datasets may include the parcel as the most fine-grained land unit, and linear features such as streets may be mapped only as centerlines. Campuses, even those set in rural regions, generally have an urban density and pattern on the campus proper. In this context, correct outlines of buildings, streets, sidewalks, and other features need to be accurately mapped. Because of the density of urbanization, physical planning and management GIS applications tend less toward natural or generalized land characteristics and more toward urban form, programmatic needs, facility characteristics, and space allocation. Utilities and related infrastructure tend to be more concentrated and complex. Within many governmental jurisdictions, much of the spatial data is tied to taxation and regulation functions of government, although some relates to assets. These two governmental functions, taxation and regulation, are generally not applicable to universities; instead, motivation for maintaining spatial data relates more strongly to asset management and institutional or physical planning. With fewer case studies and existing analytical procedures to draw on to guide applications, the boundary between GIS and traditional tools becomes blurred and new applications must be envisioned from scratch. New applications are most likely to be envisioned by staff in applied disciplines such as planning or transportation, particularly if they possess GIS skills. It is not reasonable to expect those whose sole specialty is a certain technology, such as electronic drafting, to be responsible for envisioning a new application in another discipline. Engaging an array of departments or disciplines becomes essential for envisioning GIS applications of relevance to an "urban core setting."

Interpreting the Organizational Culture

This section examines several organizational issues in the context of university GIS initiatives. The assumption is that these are not, for the most part, patterns one can consciously change, but only respond to constructively. These are the "institutional existing conditions" one must work with. In contrast, the final section of this chapter presents topics that, hopefully, are more directly under the influence of GIS professionals (or perhaps should be the focus of efforts to influence change).

Collaboration. Any GIS initiative broader than a single department hinges in large part on collaboration. Relative to GIS implementation, the existence of a culture of collaboration among departments is likely to be the most important condition an institution can be blessed with. A number of observations, which follow, give insight into why collaboration may turn out more or less successfully.

Collaboration is necessary because of the interconnectedness of a large institution. For decisions to be made in the context of the larger "whole" of the university, they must draw on information stored in various departments. With collaboration, GIS can form the foundation upon which shared spatial data are built.

Administrative-level commitment to collaboration is a valuable asset. Even if resources aren't shared and departmental boundaries are clear, a management commitment to collaboration helps enable technical staff in various departments to work together more effectively. Skill and good relations on the technical level are enhanced by good communication and common goals among administrative staff.

If a project or a system mandate is endorsed by all relevant departments and the administration, collaboration becomes easier, in part because there is greater clarity about each department's role. For example, it becomes possible for an appropriate department to take a leadership role without such effort possibly being misconstrued as inappropriate meddling or control. Mandates suggest a vision and focus, both of which are needed to move a group of departments in a consistent direction.

Many interdepartmental collaborative projects don't have a discrete budget allocation; therefore time and expenses must be drawn from individual department budgets. This may necessitate cutting into departmental priorities, which can deter collaboration. Some departments are set up to work only on billable projects; therefore collaboration is more likely if a discrete budget exists for the GIS project or system.

The degree to which information flows across departmental boundaries has an impact on the development of enterprise GIS since it is inherently built on shared and integrated data and systems. Many departments are likely to work well together, but some level of territoriality is probably inevitable; therefore consciously working to build bridges is essential. Sharing data and jointly developing projects raise issues such as deciding the location from which software applications are served, the location for data storage, who receives credit, and so on. Collaboration may be particularly fruitful between departments within different major divisions of the university, or even with departments outside of the university (for example, working with the county government on the segment of a wastewater sewer map that covers the campus).

Incompatible data and software formats may become a rationale for choosing to not participate in a collaborative effort. A university is an incredibly diverse work environment in terms of information needs, communication styles, job titles, professional disciplines, goals, and so on. It is difficult to meet everyone's needs with one system in such a diverse work environment. For example, a campus base map that is accurate to within one foot horizontally is not likely to be accurate enough to satisfy engineers, while at the same time it may be seen as overkill for many users who simply want an illustrative "you are here" type of map.

These are all valid needs that must be accommodated if a system is to serve a large enterprise. Meeting diverse needs means integrating heterogeneous data and related software systems, which adds significant complexity to the process. Without various departmental staff (or perhaps an entire subdepartment) having a mandate to work through these complexities, collaboration may suffer because it is easier for individuals to retreat to tools they're familiar with and that complete the immediate task at hand. Information technology staff may be leery of technology solutions that are centralized or top-down, or that otherwise limit their personal preference for what they believe is the most appropriate solution. This perspective is clearly more justified on the academic side of a university given the great diversity of disciplines; however, it may present barriers to collaboration within administrative units where there is a high expectation of running an efficient, integrated, businesslike operation. If a precedent exists of maps being created with computer-aided design and drafting (CADD), there may be challenges integrating GIS data and software since each technology comes with a unique data format and each was built with a particular use in mind. Such technological hurdles may be overcome relatively easily, however, when operating in a positive and collaborative context.

Technology Advocacy. The level to which technology is used and advocated throughout a university may offer a clue as to how well GIS will be received. In the days before automation, there were most likely advocates for better paper map products. Because of the technological basis for GIS, advocates must now be cultivated for technology products as well: databases, technical documentation, online interfaces, and so on. Facility-related staff are often from a design, planning, or construction trade background as opposed to a technical, GIS, or computer background. Such staff are the best source for application ideas, but many do not possess extensive technical or computing skills. Nonetheless, advocacy for technology at any level within the university is a helpful boost. Technological change, such as the World Wide Web and networking, has raised the bar of expectations in the GIS field. For example, interactive Web maps are a great asset, although their apparent ease of use may create a disconnect between the desired end products and the available resources required to create those products. Administrators, and those with technology experience in particular, should be encouraged to serve as champions for enterprise GIS.

Decision Support and Structured Analyses. The level to which structured or data-driven analyses are used as "decision support" may be an indicator of the degree to which GIS services will flourish. Such analyses inevitably combine with expert opinion in decision making; however, with a precedent of incorporating structured analyses, support for GIS may well be stronger. If conclusions tend to be determined by professional judgment alone, the role of GIS may be limited to graphically rendering solutions or decisions. This contrasts with using maps for quantitative analyses, which help formulate and inform decisions. Meaningful analyses require up-front time for project design and data assembly. Clear priorities and appropriate lead time between project requests and product delivery deadlines help support proper database design and documentation. Wherever possible, it should be reinforced that meaningful analyses require extra resources but are well worth it.

GIS Applications and Map Functions. Many people may not realize there is more to GIS maps than the final graphic image, so it is important to foster understanding of the difference between one-time "graphic maps" and map products based on GIS functionality. In a dynamic and rapidly changing work environment, where a major focus is on rendering and selling proposals, or preparing presentations and publications, tools are likely to migrate toward those excelling at graphic quality, expediency, and ease of use. In this context, GIS benefits (notably data linking, analysis, spatial accuracy, and future utility of the product) may not be highly relevant. This tradeoff—expediency and graphic quality in exchange for accuracy, continuity, analysis—should be carefully evaluated when selecting mapping tools. The idea of a map as a graphic representation of data is a relatively new concept for many people. Consequently, effort must go into demonstrating that *maps as data visualization* truly enhance current thinking on issues.

Similarly, the distinction between mapping and design should be recognized. GIS benefits are frequently tied to the ability to better understand existing conditions; if used in the context of a design project, its value is therefore likely tied to the part of the process that involves analysis of existing conditions. GIS is not intended as a design tool, so it is less relevant when the task at hand is electronic drafting or design.

Data Stewardship. The term *data steward* refers to a department or individual responsible for maintaining specific institutional data over time. Maintaining data over time is a foundation of GIS. As an organization embarks on GIS and becomes a data steward, resource allocation concerns may emerge since time and other resources need to be redirected toward work that yields no immediately visible "front counter" product or service.

Departmental Focus and Structure. The work focus of a department engaging in a GIS initiative may indicate how well the technology is accepted. For example, planning departments normally deal with a mixture of processing development projects and plan development. A focus on detailed programming, budgeting, or marketing of "project scale" work for individual buildings may generate less justification for use of GIS as a decision support tool than, for instance, planning work that looks at land use, urban patterns and systems, and environmental issues. Even though GIS can be used to produce diagrams or illustrations of proposed projects, such efforts are more often associated with graphics or CADD software. When staff responsibilities include both rendering and diagramming proposed facilities and maintaining maps of the existing campus, the facility work is likely to take precedence. One successful organizational structure for GIS services is to collocate within a technology-related subdepartment that provides traditional information technology functions, GIS services, and often space accounting.

Collegial Support. The presence of colleagues who share similar professional interests and expertise constitutes an enormous value in any work environment. Without such colleagues to commiserate with and lean on for technical support, each step of the process may draw out because one must take on the sole responsibility for addressing all key areas of a GIS operation. Collaboration is inherently easier with those who share some basis of knowledge, language, and interest. Efforts should be made to develop a "community of colleagues," be it within the department, among other campus departments, or with the larger GIS community.

In sum, assessing a university's culture in terms of the dynamics discussed here helps in developing strategies and a system focus. Wherever possible, strengths of the university's organizational culture should be exploited while buffering weaknesses.

Forms of Administrative GIS on Campus

The discussion now turns to a review of basic forms of campus GIS, organized around functional categories and scale of initiative. Functional categories address what the systems are used for, while scale of initiative relates to variables such as organizational extent, integration, and complexity of the system.

Functional Categories of GIS. Campus administrative geographical information systems cluster into several functional categories: technical reference information, public reference information, and decision support.

Technical reference information includes detailed maps showing the location and type of physical features, such as utility lines. The maps are of primary interest to planners and managers who need accurate and detailed information about the campus to do their job. The focus is more on accurately conveying technical information than producing a graphic composition. Stewardship of this information is generally housed within a facility-related unit of the university, usually a subsection of either a planning department or a facilities management or maintenance department. The section could be thought of as the "maps and records" department and normally include a combination of electronic information and archived paper drawings. Technical reference information usually emphasizes building floor plans and related space data, but it may also include campus utilities, base maps, photos and other images, future plans, layers for a variety of existing conditions, and a range of tabular information related to the physical campus. This typically includes archiving and imaging done for architectural and infrastructure "as-built" drawings.

Most often, technical reference information that has been converted to electronic form relates to buildings and has been developed with or in conjunction with CADD software. By default, CADD becomes the tool of choice for other maps and drawings since it represents the available skill set. For the most part, these departments are in the business of maintaining electronic drawings and generating electronic or hard copy versions as requested. At some universities, this electronic information has taken on new meaning and function through publication on Websites, which allows viewing, querying, and simple spatial analysis.

Observations suggest that the success of GIS within a technical reference information unit depends greatly on precedents within that unit. It may be difficult to gain general acceptance for using GIS if there is a strong precedent of CADD or no history of developing new map layers atop a base map established with a standard coordinate system. In such a case, significant GIS use is most likely to materialize only in the presence of a strong top-down mandate. Technical reference information is a highly appropriate use of GIS, but its adoption may be somewhat tenuous because it is the one GIS function most easily replaced with CADD (both CADD and GIS can be used for creating electronic drawings). Distinguishing aspects of GIS, such as database functionality, have great potential in this arena but do not currently seem to be fully recognized, perhaps because the historical precedents in this arena were static paper drawings prepared with conventional drafting techniques (the precursor of CADD).

The second general campus GIS function, *public reference information,* includes maps in which special care is given to the readability of graphics and text, with the intent of clearly communicating information about the location of features to a specific "public" audience. In essence, this is the realm of classic cartography (for example, road atlases). Applied to the campus context, this includes products such as directory maps for finding one's way, official campus Web maps, visitor parking maps, and many others. As with technical reference information, public reference information is typically created in a mixed software environment. Because of the need for exceptional graphic quality and the relative ease of use of graphics software, public reference maps are often created with graphics programs such as CorelDraw and Adobe Illustrator. At the University of Arizona graphic maps of this kind are increasingly derived from GIS base information; however, in the past many departments maintained their own graphic base maps. They add complexity when attempting to standardize on one consistent source of geographic information because the origin of the spatial information is frequently unknown. In addition, graphic maps generally are not scalable, and since their features aren't normally in a standard coordinate system they cannot be easily overlaid with GIS features. As GIS software has evolved, graphic capabilities are being enhanced, making it more viable for high-quality cartographic output.

The third function, *decision support,* involves use for planning, analysis, and research and is most frequently found in facility-related divisions, but also to some degree in institutional research departments. There are great untapped uses here for parking, space management, and infrastructure management, but the predominant use has been in campus planning. One such example at the University of Arizona involved address-matching residential locations of the faculty, students, and staff so as to create a point map showing the distribution of the entire campus community. The purpose was to use this information in planning park-and-ride lots and evaluating shuttle routes.

Because GIS software emphasizes analysis and data manipulation more so than CADD does, it is likely to be more highly valued for such applications. CADD software has recently been adapted to include some GIS functions, although it appears that this software is still used primarily for its exceptional capabilities in editing electronic drawings.

It should be noted that all three functional types of GIS have existed on campuses as part of what is termed a "request and delivery system" (Cardenas, 1998). In a system of this sort, a GIS is limited to a cycle of maps being requested and delivered. As a result, new map layers and most other GIS work become a consequence of the request-delivery system; without new map requests, the production of mapping halts, and data input recedes. Expanding the GIS to other departments, given this system, is not possible since other departments are unlikely to possess necessary skills; the GIS is viewed only as a map production tool, and others are content with continuing to request maps from the GIS department. Cardenas recommends embedding highly customized and simplified GIS applications within nonspatial software in departments throughout the organization, thereby making the GIS an influence on planning and decision making throughout the organization. The focus of a GIS department should shift from map request and delivery to the development of imbedded applications.

Daniel (1995) notes that it is all too easy to pave over old cow paths, meaning that automation efforts often simply provide an electronic means to do the same business tasks we have always done. In other words, the challenge is not in selecting GIS as the technology of choice for current mapping tasks but to determine what new value more powerful and intelligent mapping systems bring to addressing business problems.

These ideas point to the need to move away from thinking that the only difference between digital maps and paper maps is the medium. We should envision how digital maps, and GIS software, can help us work more effectively and in ways we never imagined could relate to spatial data. An example, found on some campuses, is an imbedded application allowing access to a user-restricted Web interface where building occupants can query and update their own space data online.

Scale of GIS. On the basis of observations of many campus GIS initiatives, it appears they scale along several continuums: the number of people and departments involved, the use of off-the-shelf software as opposed to development of custom tools and interfaces, the amount of central mandate, and the level of integration between departments. Three basic scales of campus GIS are discussed here: a departmental GIS, a system made up of a loose confederation of departments, and a fully integrated and developed enterprise system.

Departmental GIS. A departmental GIS involves one or more staff members in a university department using off-the-shelf GIS software, often with all the GIS data they use residing on personal computers. Data may reside on a departmental server computer, but one person typically uses it at a time for discrete mapping projects. This model is the simplest for getting a basic system up and rolling, although it is more complex than it may appear. To begin any form of GIS, a base layer must be created, which requires the same amount of work regardless of the size of the department. Without resources to contract for services, a small department needs considerable time to create a functional GIS. Any GIS organization includes a number of essential roles, and in a small organization one person or several people meet all these roles. Conversely, larger GIS organizations may have several people fulfilling a single role (for example, database management). Most campus GIS efforts appear to have begun as a departmental initiative, followed by incremental steps toward the second type.

Loose Confederation of Departments. With this model, several departments use GIS and CADD software and attempt to share data, usually across a network or copied onto removable media for transfer. It is unlikely such departments access the same datasets on any one server or update and access information using a common customized software interface. One benefit of this model is that departments are less likely to recreate the same layers if they can obtain them from another source. Basic coordination and agreement are required on the use of standard base layers, data stewardship, software formats, data conversion methods, and so on. Most campus initiatives reviewed here fall into this category.

Fully Integrated Enterprise System. The third scale, the enterprise, may vary in its level of centralization, the degree of software customization, the number of departments it serves, and the application types it provides. The software environment could be thought of as a toolbox of automated and interconnected functions, customized for specific users who rely on shared datasets. This sort of system affords the greatest integrated functionality for a range of campus users, but it is also the most resource-intensive initially. The best examples of enterprise automation within a university tend to be the systems that manage financial, personnel, and student data. For example, many universities have online systems for payroll processing, which allow every department to submit payroll data from a departmental computer. Review of other campuses suggests that systems that might be considered enterprise GIS occur primarily for technical reference information, with most focusing on space databases and building floor plans.

Some universities have created a degree of enterprise system for the public reference information function by way of Websites offering a number of map products and some limited query functions. Most of these systems appear to knit together standard software tools (AutoCad, ArcMap) into a custom system, although some prepackaged systems are in use, such as Archibus. Efforts at automating maintenance, repair, and central plant operations exist, but they appear to be systems unto themselves as opposed to being integrated within a larger family of enterprise information systems. GIS functions could be integrated with automated utility and building control systems, but little evidence of these functions has turned up. Examples of enterprise GIS are, at best, only formative at this point and exist for a limited functionality such as space management. A number of universities anticipate that planned interactive Web maps will become the foundation for the enterprise GIS at their campus.

Key Strategic Issues for Enterprise GIS

A number of fundamental organizational issues must be addressed if a system is to evolve beyond a single departmental level. What follows is not intended as a catchall menu for creating a GIS, but rather ideas on moving toward an integrated, enterprise GIS. Further, these are topics over which GIS professionals should have some control and are likely to be the most fruitful focus of their energy.

Funding. A key strategy in initiating enterprisewide GIS is to develop administrative understanding of GIS benefits. This understanding is a foundation for future funding. Administrators appropriately focus on high-level strategic and budget issues; therefore time available for presenting technology issues must be carefully used if key objectives are to be achieved. It is possible to fund a GIS from individual departmental budgets, although it is unlikely for a truly integrated system to emerge under this circumstance. A dedicated budget for a GIS department or initiative probably requires central funding, which raises the issue of how to organize the GIS department within the larger organization. In some instances, a separate unit is created that reports to the administrator in charge of participating departments.

A misperception can occur—that once the software is purchased, one has a functional GIS. Success in GIS is correlated with continuous allocation of resources, thereby allowing periodic data updates, new software and hardware, application development, staffing, and training.

Interfaces and Applications. Simple online interfaces, permitting access to basic information people need to do their job, are likely to be the most valuable application in leveraging limited start-up dollars. In addition, highly customized and specialized applications are needed if non-GIS users are expected to produce GIS tasks, analyses, or maps. Casual users probably will not directly use out-of-the-box GIS software. Discussion about proposed interfaces should emphasize that they can be password-protected. With homeland security concerns, some campuses have become leery of putting geographic data online. A range of institution-specific applications will emerge when a "GIS assessment" study of the university is conducted.

Establishing Agreements: Roles and Standards. The role of each participating department should be established, accepted, and clearly documented. It is important that responsible parties be designated, both to ensure tasks are completed and to minimize duplication of effort. Many university departments are potential users of GIS products; however, there seems to be little practical ability to create or maintain systems that generate these products. Any data element being shared should be fully documented (data field characteristics, naming conventions, standards for drawing elements, frequency of update, source, and other metadata). Software standards may also be useful, depending on the context.

A Focus on the Data. Acquiring base data to begin building the system probably represents the largest one-time investment of resources overall. The "data foundation" is a campus base map built upon an established coordinate system. Creation of a base map and seeing that all subsequent mapping is referenced to it are the two fundamental steps. Designating responsible data stewards and establishing data standards are foundations from which to build an enterprise system.

Demonstration Projects and Small Successes. Most GIS initiatives are built incrementally. As such, it is necessary to show small successes to convince decision makers of the value of the GIS system and to furnish additional resources where necessary. It is advisable to prioritize key demonstration projects that are achievable given current timelines, human resources, and data. The projects should revolve around meeting a customer need, which creates advocacy and ongoing demand for GIS services. Although simple demonstration projects may be relatively inexpensive and achievable in little time, plans should be made for custom online applications so that GIS function becomes imbedded in decision support areas throughout the institution. Keep in mind the need to continuously demonstrate technical progress, while not losing sight of making plans, conducting outreach, and working through administrative channels for support. Administrators who fund projects want to see evidence of a product.

Outreach and Education. Outreach to other departments and entities is important in creating understanding and support for a GIS initiative and for building bridges. In an environment with minimal staffing, outreach is difficult, but, even so it should be a continuous activity. Some predictable types of outreach presentation are to the campus community, to decision-making bodies, to conferences, to student groups, and to external government entities. Software training classes are a good strategy to persuade departments to make initial, tentative commitments to GIS. Other outreach includes preparation of "white papers" on issues and possibilities, and meetings about organizational and technical issues such as data conversion.

Documentation. GIS documentation is essential in maintaining technical organization and also an effective tool for communicating what has been accomplished, what the possibilities are, and the complexity of work involved. For an enterprise system, one where multiple data sources and software systems are being integrated, high-quality, accessible documentation is essential for the system and for people to communicate about the system. Graphical documentation of data models can be useful for assessing and building linkages with other campus data systems, and it can illustrate the breadth of information to be considered "geographic information."

Persistence. Creating an enterprise system takes time. The most advanced campus GIS initiatives, which generally have not moved far beyond the "loose confederation of departments" model, have been in progress for ten to fifteen years, and some even longer. A focused top-down mandate with associated funding could bring about an enterprise system relatively quickly, but in this context relatively quickly may mean three to five years. Most systems proceed incrementally; thus, patience and persistence are essential.

Ideally, all these factors, and surely others, are fully explored at the beginning of an initiative in a system assessment study. After this, system specifications are developed, followed by an implementation program, perhaps done by way of a series of prototype projects. Many references are available that describe the basic components and steps used in these documents.

Conclusion

The university campus is a unique environment for GIS. Because GIS is a relatively new area of application, few precedents or standards exist, thereby presenting challenges for new initiatives seeking guidance. At the same time, there is tremendous untapped potential for GIS applications.

This chapter explores administrative GIS and the potential for creating an enterprise system—from an organizational and strategic standpoint more than from a purely technical one. One realization that resulted from this author's research is that there are numerous common issues and applications found among campus GIS initiatives. Organizational dynamics were reviewed. Collaboration, for example, was examined as a fundamental requirement for enterprise GIS because it derives from a philosophy of integrating and sharing information.

The campus planning department is often the source of an initial vision for a campus GIS. Ironically, as a small department, it may not have the resources to take an initiative to the enterprise level. A major unit, one that is responsible for maintaining a large portion of the physical campus, may have a better pool of resources to draw on. At the same time, many successful campus GIS units are housed within a planning department. The key to this approach appears to be establishing a discrete subdepartment that has been appropriately funded to serve not just the immediate department but the entire campus as well, and to maintain all facility-related information resources in one integrated unit. Regardless of the department GIS is housed in, the most important strategy may be to direct resources to where the vision is.

An emerging vision for enterprise GIS focuses on customized Web interfaces allowing appropriate faculty, students, and staff to access all the information they need to do their job, regardless of where it is stored. This also includes the ability to enter and update data where appropriate, along with online tools that allow processing of data or performance of analysis unique to their job. The Web interfaces are linked to a spatial data warehouse, which is the source of most data.

Functionally, the sky is the limit. Virtually any department can become part of the system, assuming it has a use for spatially referenced data. The vision is that fundamentally different approaches to defining and solving problems will emerge. Decision making becomes more distributed throughout the organization, and conventional obstacles to responding quickly, accurately, and thoughtfully no longer exist.

References

Cardenas, H. D. "The Integration of Geographic Information Systems in Municipal Governments." In *Proceedings* of the 1998 ESRI User Conference, 1998. San Diego, Calif., July 27–31.

Daniel, L. "Expanding the Business Geographics Agenda." *Business Geographics,* Sept./Oct. 1995, p. 6.

B. GRANT MCCORMICK is a GIS coordinator in the Office of Campus and Facilities Planning at the University of Arizona.

7

In a time of decreasing institutional resources, the ability to refine and target donor campaigns is more critical than ever. This chapter demonstrates how GIS can be used to better understand where alumni are located and, further, where to focus future campaign activity.

Using GIS in Alumni Giving and Institutional Advancement

Daniel D. Jardine

The Office of Institutional Research at Binghamton University (SUNY) analyzed donating patterns and behaviors of BU alumni from 1992 to 2002. The analysis produced descriptive information on alumni giving patterns and characteristics and helped the university to develop predictive models of alumni giving that allow more targeted solicitations.

The database used in the analysis includes more than seventy-five thousand alumni records. The vast majority of these records contain address-related data, particularly the zip code of the individual. With the zip code, the alumni data could be geocoded or assigned coordinates to be placed on a digital map in a geographic information system (GIS). Using a GIS can greatly enhance the ability to better visualize alumni data to improve the planning efforts of future campaigns.

This chapter contains a series of national- and state-level maps. The maps illustrate where BU alumni live, where donors live, and how much money has been donated by location (state and county). The alumni office uses these maps to better understand where alumni are located and consequently where to focus future campaign activity.

An example of a map recently requested by the public relations office is included. This map displays the number of alumni and current students, residing in key congressional districts, who serve on appropriations committees. These committees appropriate research dollars at the federal level. GIS can be used to illustrate the magnitude of the presence of alumni and students in these key districts.

NEW DIRECTIONS FOR INSTITUTIONAL RESEARCH, no. 120, Winter 2003 © Wiley Periodicals, Inc.

Alumni Data

The university alumni database contained 75,973 records as of spring 2002. Since the database is used for mailing purposes, the alumni office keeps address information as up to date as possible. The number of records containing a U.S. zip code was 63,385, or 83.4 percent. The number of those records matching a U.S. zip code boundary in the zip code digital map reference file was 62,219, representing a match rate of 98.16 percent. For New York State, there were 38,120 records. Of these, 37,660 matched the zip code boundary file. This represents a match rate of 98.79 percent. For the purposes of mapping alums by state, the State field was used. This resulted in a total of 63,302 alumni records matching a state. The next section describes the use of the geocoding method to analyze the distribution of alumni by state.

Alumni by State

Figure 7.1 illustrates the distribution of Binghamton alumni using the 63,302 alumni records that contain the state in which the alum resides. The map clearly shows that the majority of BU alumni (38,120) are currently living in New York State. This represents 60 percent of all alumni.

It is interesting to note that approximately 95 percent of BU students come from New York State. New Jersey is the state with the second largest alumni population (3,716, or 6 percent). There are eight additional states with more than 1,000 alumni: California (2,612), Pennsylvania (2,111), Florida (1,921), Massachusetts (1,795), Connecticut (1,437), Virginia

Figure 7.1. Number of Alumni by State, Spring 2002

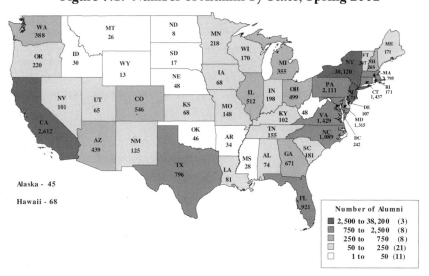

(1,429), Maryland (1,315), and North Carolina (1,089). With the exception of California and Florida, these states are contiguous and run from Massachusetts down to North Carolina.

New York State Alumni by County

Figure 7.2 illustrates the distribution of alumni living in New York State by county. The county with the largest number of alumni is Broome, which is the location of the university.

More than seven thousand BU alumni are currently living in Broome County. This is roughly 19 percent of all New York State alumni. Tioga County, located adjacent to Broome on the west, has nearly one thousand alumni. Broome and Tioga counties make up a standard metropolitan area, and they are closely linked in terms of economic and social activity.

Counties that are located in the New York City metropolitan area also have a large number of BU alumni living there. In fact, the New York metropolitan area contains more than twenty thousand alumni, or 54 percent of New York State alumni. In other words, well over half of the BU alumni living in New York State live in the New York City metro area.

For many years, the percentage of students who come from the New York metro area has been roughly 65 percent. The map reveals that Nassau County on Long Island is home to 4,320 BU alumni, or 11.5 percent of New

Figure 7.2. Number of Alumni in New York by County, Spring 2002

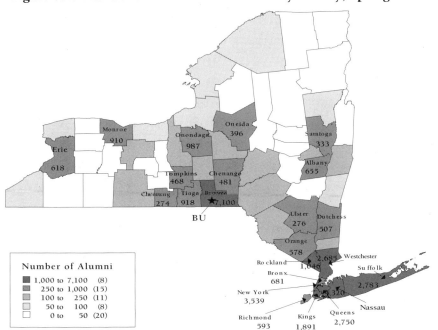

York alumni. This is followed by New York County, more commonly referred to as Manhattan, with 3,539 alums (9.4 percent); Suffolk County with 2,783 (7.4 percent); Queens (2,750, 7.3 percent); Westchester (2,685, 7.1 percent); Kings, more commonly known as Brooklyn (1,819, 4.8 percent); and Rockland with 1,046, or 2.8 percent of New York alumni.

Counties outside of metropolitan New York with a relatively large number of BU alumni are located in other metropolitan areas: Onondaga (Syracuse metro area) with 987, Monroe (Rochester metro area) with 910, Albany County (Albany metro area) with 655, and Erie County (Buffalo metro area) with 618.

Donors by State

Figure 7.3 illustrates the number of alums who donated to BU between 1992 and 2002. A total of 24,941 alumni gave a donation to the university between 1992 and 2002, a participation rate of 39.4 percent. Not surprisingly, the states with the largest number of alumni also have the largest number of donors. New York State had 13,714 alumni who gave a donation in that time frame.

Other states with a relatively large number of donors are New Jersey with 1,977, California with 1,020, Pennsylvania with 862, Massachusetts with 826, Florida with 792, Connecticut with 786, Virginia with 648, and Maryland with 629. No other state has more than 500 donors.

Figure 7.3. Number of Alumni Who Donated by State, 1992–2002

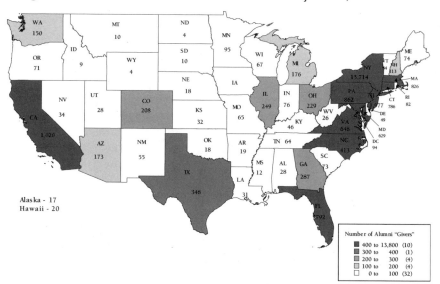

Participation Rates by State

Figure 7.4 illustrates the percentage of alumni who gave a donation between 1992 and 2002 by state. New York State had a participation rate of 36 percent.

There are seven states with a participation rate of 50 percent or greater: South Dakota (59 percent), Arkansas (56 percent), Connecticut (55 percent), West Virginia (54 percent), New Jersey (53 percent), North Dakota (50 percent), and Michigan (50 percent). Unfortunately for the university, most of these states have a very low number of alumni; thus the high participation rate is relatively meaningless. There are a few states, however, that have *both* a relatively large number of alumni and relatively high participation rate: New Jersey (3,716 alums, giving rate of 53 percent), Connecticut (1,437 and 55 percent), Maryland (1,315 and 48 percent), and Massachusetts (1,795 and 46 percent).

Donors by County in New York State

Figure 7.5 illustrates the distribution of donors by county across New York State. Of the 37,660 BU alumni living in New York State, 13,714 gave a donation to the university between 1992 and 2002, for a participation rate of 36 percent.

There are five counties that contain more than 1,000 alumni who gave to the university in this time period. Not surprisingly, Broome County,

Figure 7.4. Percentage of Alumni Who Donated by State, 1992–2002

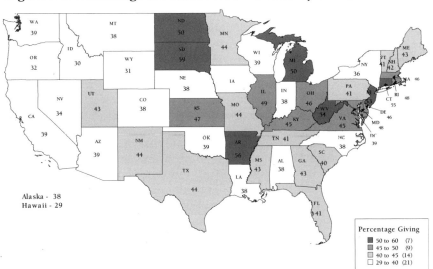

Figure 7.5. Number of Alumni Who Donated in New York by County, 1992–2002

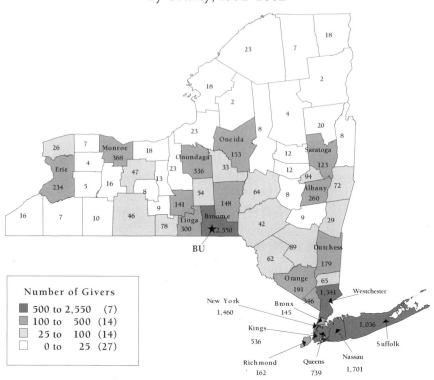

where the university is located and the county with the largest alumni population, has the largest number of donors, with 2,550. The four other counties with more than 1,000 donors are all in the New York City metro area: Nassau (1,701), New York/Manhattan (1,460), Westchester (1,341), and Suffolk with 1,036 donors.

Participation Rates by County in New York State

Figure 7.6 shows the participation rate by county for New York State. The county with the highest participation rate, by a large margin, is Westchester.

Westchester also contains a large number of alumni, with nearly twenty-seven hundred. Fully one-half of alumni living in Westchester made a donation to the university between 1992 and 2002. Other counties with relatively high participation rates include Schenectady with 44 percent, Greene with 43 percent, Columbia with 42 percent, New York (Manhattan) with 41 percent, Albany with 40 percent, and Monroe with 40 percent. As Figure 7.6 illustrates, four of these top counties are concentrated in and

Figure 7.6. Percentage of Alumni Who Donated in New York by County, 1992–2002

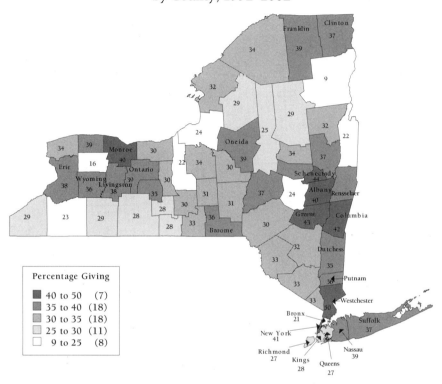

Percentage Giving

■	40 to 50	(7)
■	35 to 40	(18)
▨	30 to 35	(18)
▢	25 to 30	(11)
□	9 to 25	(8)

around the Albany metro area. One of the benefits of using a GIS to map participation rate—as opposed to looking at a participation rate table, where the locations of some counties may not be known to the reader of the table—is that clusters of counties become quite apparent.

Total Donations by State Between 1992 and 2002

Figure 7.7 indicates the total dollar amount donated by alumni for each state between 1992 and 2002. The amounts range from a low of $617 donated from alumni living in North Dakota to a high of more than $7 million donated from alumni living in New York State.

Other states with large contribution totals include New Jersey, with nearly $1.3 million; Washington, with nearly $800,000; Massachusetts, with nearly $750,000; and Washington, D.C., and Connecticut, with more than $500,000 each.

Figure 7.7. Total Alumni Donations by State, 1992–2002

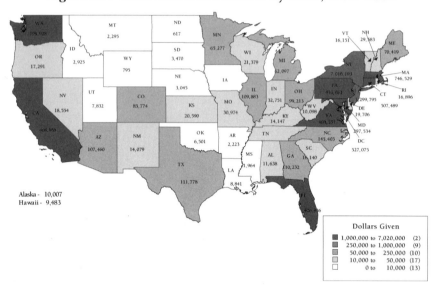

Total Donations by County in New York State Between 1992 and 2002

Figure 7.8 displays the total alumni donations by county in New York State between 1992 and 2002. There are three counties with alumni donations topping the $1 million mark: New York (Manhattan), Broome (BU's location), and Westchester. Again, other metropolitan areas such as Erie County (Buffalo), Monroe (Rochester), and Albany (Albany) also show relatively large total amounts donated.

Average Total Donations by State and County

Figure 7.9 shows the average total dollar amount given by state between 1992 and 2002. For example, if a donor gave $50 a year for each of the ten years between 1992 and 2002, his or her total donation would be $500. The average ten-year total given by all donors was $543.

Looking at the map, one sees many states near the $500 level. In fact, there are fifteen states with an average between $400 and $600 and thirty-seven states between $250 and $750. Two locations stand out with averages substantially higher than all others: Washington, D.C. ($5,607), and Washington State ($5,333).

Figure 7.10 shows the average total amount given by county for New York State. The average total given from 1992 to 2002 by donors who live in New York State was $512, slightly below the overall average of $543.

Figure 7.8. Total Alumni Donations in New York by County, 1992–2002

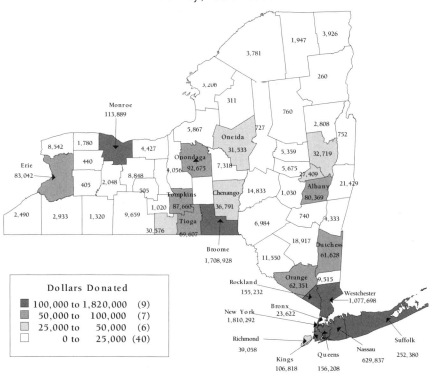

When the data are disaggregated to the county level, however, only four counties had an average total donation of more than $500. The highest average total came from New York County (Manhattan), with a ten-year average of $1,240 per donor. Westchester County had an average of just over $800, while Broome and Tompkins counties averaged $670 and $622 per donor, respectively.

Figure 7.11 is an example of overlaying one variable on another. Here, we take the average amount given, represented as a graduated symbol layer (the larger the symbol, the higher the average donation) overlaid on a map illustrating median income. This gives a glimpse of the relationship, if any, between average donation and median income at the state level.

One observation gleaned from viewing this map is that the seven states west of the Mississippi River in the lowest income quartile (the lightest shading) all have relatively low average donations. The average total donations range from $117 in Arkansas to $361 in Oklahoma. (Again, the average total donation was $543.) Perhaps the combination of lower income and distance from Binghamton University contributes to the relatively low level of donations from these states.

Figure 7.9. Average Total Donation by State, 1992–2001

Figure 7.10. Average Total Donation in New York by County, 1992–2001

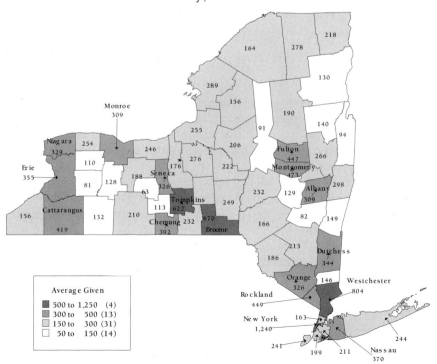

Figure 7.11. Median Income and Total Alumni Donations by State, 1992–2001

Figure 7.12 shows the average total donation overlaid on median income by county for New York State. Of the four counties with the largest average donations, only Westchester falls into the highest median income category.

New York (Manhattan) is in the second median income group, while Broome and Tompkins are in the third income group. Of the counties in the highest income category, Rockland is the one with the highest average donation, at $449. It appears that this map does not reveal any obvious relationships between median income and average donations at the county level for New York State.

Alumni and Students in Key Congressional Districts

Aside from mapping alumni giving patterns, GIS can be a valuable tool in illustrating the presence of alumni and students in key congressional districts. Figure 7.13 was created specifically for a visit to key congressional representatives serving on appropriations committees that deal with federal funding for research.

A top university goal is to increase funding for research. The map shows that in each district, there are more than one thousand alumni. One district has more than two thousand alumni of the university. The map

Figure 7.12. Average Total Alumni Donation in New York by County, 1992–2001

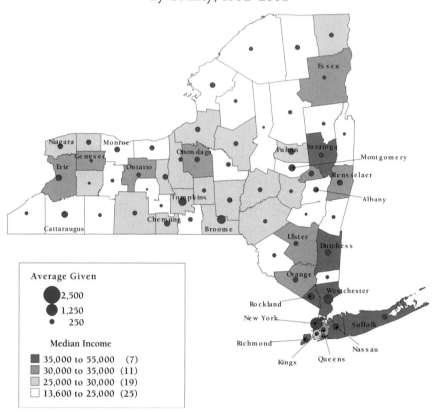

also illustrates the large number of current students who come from these districts. This is a helpful tool in making the case that the university has an important presence in each of the districts.

Conclusion

GIS can be a valuable tool in mapping institutional data of many kinds, particularly alumni data. As the maps presented in this chapter illustrate, location is a critical component when planning a university's fundraising campaigns. Maps can increase the level of understanding of where alumni live, areas with a large number or percentage of donors, and areas with a large number of likely donors. In a time of decreasing institutional resources, the ability to refine and target donor campaigns is more critical

Figure 7.13. Number of Alumni and Current Students in Selected Congressional Districts of New York

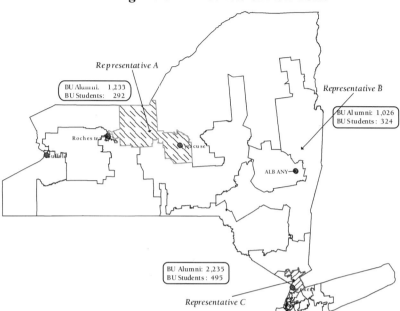

than ever. Binghamton's model demonstrates how GIS can be used to target donor campaigns and help illustrate the university's presence to key people involved in appropriations for research and other federal funding.

DANIEL D. JARDINE is a research analyst in the Office of Institutional Research at Binghamton University in Binghamton, N.Y.

8

This chapter discusses useful resources for learning about GIS analysis, software packages, and sources of GIS data.

Bringing GIS Analysis to Institutional Research: Resources for Practitioners

Daniel Teodorescu

Although GIS is relatively new to the world of institutional research, the chapters in this volume illustrate a range of possible applications in higher education administration. This tool becomes necessary as an increasing quantity of data in our research projects includes a spatial component. (One should note that some of the data that are critical to academic administration have no spatial component and cannot be used by a GIS. Typical examples of IR datasets in this category are faculty workload and student credit analysis.) Although statistical methods can help us answer questions of *what* (descriptive statistics), *how*, and *why* (inferential statistics), they are not very good at answering questions of *where*. It is only through GIS software that we can easily examine overlapping layers of information. This final chapter presents a few concluding remarks and discusses a series of resources that institutional researchers could use to learn more about this.

First, as several applications in this volume suggest, GIS can offer an edge in a competitive environment. In light of recent economic downturns and reduced government funding, targeting recruitment and solicitation efforts will become even more important. Mora's Chapter Two illustrates how admissions personnel can use GIS to better understand where potential students live in relation to the campus and competing institutions and how this information can be used to target marketing efforts.

Likewise, GIS can help universities pursue funding from donors and alumni clubs. As Jardine demonstrates in Chapter Seven, by linking alumni data with census-based demographic information, GIS can assist alumni offices in locating, cultivating, and soliciting those who are more likely

NEW DIRECTIONS FOR INSTITUTIONAL RESEARCH, no. 120, Winter 2003 © Wiley Periodicals, Inc.

to give to their alma mater. GIS can also be used to predict and visualize demographic changes in neighboring counties and states and thus allow a college or university to better prepare to meet such changes.

The second concluding remark is that although maps take up a large part of this volume, the reader should not be left with the impression that GIS is nothing more than a "mapping" tool. Indeed, GIS can be used as an effective presentation tool when communicating with top administrators on campus. But the most important benefits are evident only when one uses such a tool for analysis by asking what-if questions, exploring data for geographic patterns, integrating with other information systems, or confirming a hypothesis. GIS tools, for example, have a unique but often underexplored ability to examine data for geographical correlations. In many government agencies, GIS is already a part of decision support systems for top policy makers. If it is integrated with student, faculty, and alumni databases, it can become a valuable part of the executive information systems that institutional researchers often design for academic leaders. The ability to organize and display data in overlapping layers is a particularly useful functionality that can enrich both presentation and analysis in such systems.

As these concluding remarks suggest, GIS could be a welcome complementary set of skills if added to the statistical expertise that most IR professionals already possess. Unfortunately, these skills are not taught in social science graduate programs and most often have to be learned through professional development. To promote such learning, here is a discussion of useful resources for learning about GIS analysis, software packages, and sources of GIS data.

Introductory Books on GIS

Korte, G. *The GIS Book.* Santa Fe, N.M.: OnWord Press, 2000.
This is a practical guide written for the practitioner charged with bringing GIS into an organization. It offers an overview of the GIS industry and software, including some sample applications; discusses the criteria needed to successfully select and implement a GIS; and gives pointers to a variety of available resources for information and training. New sections in this edition cover GIS data sources, GIS on the Internet, and maintaining a GIS database.

Lo, C. P., and Yeung, A.K.W. *Concepts and Techniques of Geographic Information Systems.* Upper Saddle River, N.J.: Prentice Hall Series in Geographic Information Science, 2002.
This book approaches the study of GIS from the broader context of information technology. It covers the concepts and techniques pertaining to every stage of the systems development life cycle of GIS and its applications in various areas of spatial problem solving and decision making. Chapters include an introduction, maps and GIS, digital representation of

geographic data, data quality and data standards, raster-based GIS data processing, vector-based GIS data processing, visualization of geographic information and generation of information products, remote sensing and GIS integration, digital terrain modeling, spatial analysis and modeling, GIS implementation and project management, GIS issues and prospects, and Internet resources for GIS.

Mitchell, A. *The ESRI Guide to GIS Analysis.* Vol. 1: *Geographic Patterns and Relationships.* Redlands, Calif.: Environmental Systems Research Institute Press, 1999.
    First in a two-part series, this volume focuses on six of the most common geographic analysis tasks: mapping where things are, mapping the most and least, mapping density, finding what is inside, finding what is nearby, and mapping change. Using examples from various industries and applications, the author explains all stages in GIS analysis: narrowing down a problem to its essential element, choosing one analytical method from among several options, using software tools accurately, analyzing the results, and communicating the results to an audience.

Longley, P. A., Goodchild, M. F., Maguire, D. J., and Rhind, D. W. *Geographic Information Systems and Science.* New York: Wiley, 2001.
    This book covers topics such as GIS science, GIS practice, GIS management, the impact of Internet GIS, and GIS implementation. To demonstrate the interdisciplinary nature of GIS, the authors present a range of case studies describing implementation issues in a clear and nontechnical way. They also use several online educational resources, most notably the ESRI Virtual Campus.

Books on GIS in Public Policy

Greene, R. W. *GIS in Public Policy: Using Geographic Information for More Effective Government.* Redlands, Calif.: Environmental Systems Research Institute Press, 2000.
    The examples presented by Greene show how GIS is carrying out public policy decisions reached by voters or by their elected representatives. The cases demonstrate the versatility of the GIS technology—how it can serve across the whole range of public policy concerns—and include examples of GIS being used for monitoring the environment, redistricting, allocating tax moneys, public policy planning, and education.

O'Looney, J. *Beyond Maps: GIS and Decision Making in Local Government.* Redlands, Calif.: Environmental Systems Research Institute Press, 2000.
    This book shows how GIS technology supports decision making, how local governments are using GIS to integrate management and planning, and what issues affect the impact of GIS technology on democratic values.

Covering a spectrum of applications for local governments, the book addresses key implementation issues and common pitfalls that every city and government needs to address.

Books on Managing GIS Projects

Reeve, D. and Petch, J. *GIS, Organisations and People: A Socio-Technical Approach.* London, England: Taylor & Francis, 2000.

This book's main premise is that a GIS project should not be seen only as a technical exercise. Projects have social and organizational contexts that must be taken into account if they are to succeed. The book overviews the human side aspects of GIS, both individual and organizational.

Obermeyer, N. J., and Pinto, J. K. *Managing Geographic Information Systems.* New York: Guilford Publications, 1994.

This book is a practical guide to the key organizational and behavioral dynamics that can determine successful implementation and use of GIS. Using management and social science theory, Obermeyer and Pinto present an ideal reference for managers, analysts, and project managers working with GIS.

GIS Software

GIS software provides the functions and tools needed to store, analyze, and display spatial information. The main components of any GIS software are tools for entering and manipulating geographic information such as addresses or political boundaries, a database management system (DBMS), tools for creating intelligent digital maps that can be analyzed, query for more information, or print for presentation, and an easy-to-use graphical user interface (GUI). Many GIS programs now offer the ability to make interactive maps available from a Web server. ESRI recently delivered a new Internet Map Server for ArcView, and MapInfo has enhanced and repackaged its Web server as MapXtreme. ArcView and MapInfo are the world's most popular desktop mapping and GIS software. Both feature geographic data visualization; query, analysis, and integration capabilities; and the ability to create and edit geographic data. Both ArcView and MapInfo are reasonably easy to learn. Both are designed as a client-server application for network use and offer clients not just for Windows but also for Macintosh and Unix. Like MapInfo, ArcView offers an optional spatial database engine that makes it possible to integrate geographic data with high-end relational databases.

Although MapInfo tends to be more popular in the business community, ArcView is widely used in government organizations. Indeed, ESRI's products have become standard tools for federal agencies, including the U.S. Geological Survey, the Bureau of Land Management, the Environmental Protection Agency, and the U.S. Forest Service. MapInfo also seems to appeal more to novice GISers, whereas people who have been around GIS for some time are more likely to become ArcView users.

In addition to these two widely used products, there are many others that are simple to use and are freely available (generally the user receives a time-stamped evaluation copy or a limited-functionality release). These products enable users to view, query, and manipulate data in numerous data formats.

One such product, CartoMap (http://www.cartoworld.com/products/cartomap.htm), is a free map-viewing Windows-compatible software program produced by CartoWorld. This map viewer supports both MapInfo Mif/Mid and ArcView Shape formats and gives users the ability to pan, zoom, query, and order layers. Many other free or inexpensive map viewers can be downloaded at the Geocommunity Web site (http://software.geocomm.com/viewers).

Data Sources

A GIS can use data from a range of proprietary and standard map and graphics file formats, images, CAD files, spreadsheets, relational databases, and many more sources. Data is free or fee-based and can come from commercial, nonprofit, educational, and governmental sources; other GIS software users; and even one's own institution. Since most GIS projects in institutional research involve the use of demographic data, the data sources given here might prove valuable.

Census Data. GIS users can download for free Census 2000 TIGER/Line data in shapefile format for an area of interest. Users can choose multiple data layers for a single county or a single data layer for multiple counties and analyze them using GIS software. The TIGER/Line files were created from the Census Bureau's TIGER (Topologically Integrated Geographic Encoding and Referencing) database of selected geographic and cartographic information. TIGER was developed at the Census Bureau to support the mapping and related geographic activities required by the decennial and economic censuses and sample survey programs. The files contain data about these features:

- Line features: roads, railroads, hydrography, and transportation and utility lines
- Boundary features: statistical (for example, census tracts and blocks), government (places and counties), and administrative (congressional and school districts)
- Landmark features: point (schools and churches), area (parks and cemeteries), and key geographic locations (apartment buildings and factories)

The demographic layers are made up of a subset of commonly used U.S. Census summary attributes: average household size, family households, married-couple family with children under eighteen years, and so on. Since the TIGER/line files are in ASCII text format only, users are responsible for converting or translating the files into a format used by their specific software package.

The Geography Network. Managed and maintained by ESRI, the Geography Network (http://www.geographynetwork.com) is a global network of geographic information users and providers. It makes available the infrastructure needed to support the sharing of geographic information among data providers, service providers, and users around the world. Through the Geography Network, one can access many types of geographic content, including dynamic maps, downloadable data, and more advanced Web services. Users can register and publish data and metadata to the portal. The data sets used in dynamic data and maps are stored and maintained by the data publisher or its hosting service provider. The main advantage of using this service is that in making a request to a dynamic data and maps service, one can access the most current data available without having to store any of that data on the system or maintain it over time.

Geodata.gov. Geodata.gov (http://www.geodata.gov) is a recently launched Web-based portal for one-stop access to maps, data, and other geospatial services; it aims at simplifying the ability of all levels of government and citizens to find geospatial data and learn more about geospatial projects under way. Geodata.gov is part of the Geospatial One-Stop initiative, one of the twenty-four OMB (Office of Management and Budget) electronic-government initiatives that will enhance government efficiency. Users first launch a Web-based map viewer, where they can add various layers (national land cover dataset, shaded relief, states, counties, interstates, streams, waterbodies, and so forth). Next they set their search criteria according to content type, such as map service, geographic datasets, activities, data theme, keywords, or date ranges. The results of the search are displayed along with metadata and if appropriate a map. Users have the option of saving a search; this way the search can be saved in a geodata.gov user profile and the person will be notified when additional data updates are added for that search region. As with the Geography Network, users can register and publish data and maps to the portal.

GeoCommunity Data Depot. The GeoCommunity Data Depot (http://data.geocomm.com) houses a variety of data in support of the GIS industry. The majority of the data have been downloaded from a range of GIS Websites located on the Internet. There are also value-added data where the staff have performed some translation, attribution, analysis, or other data-enhancing operations. There is in addition a list of free nationwide or state data that can be downloaded and used in GIS analysis.

Education and Training

For those who would like to learn how to use GIS, some form of education or technical training is highly recommended. Since GIS applies a scientific process to the tasks for which it is used, good training better equips one to understand the process and apply it properly. GIS educational programs range from vocational courses that emphasize basic GIS education and skills to more traditional academic courses that provide a strong mathematical and scientific foundation for GIS. Essentially, there are five main methods to learn GIS: formal university postgraduate degree programs, instructor-led courses offered by software vendors, web-based courses, GIS conferences, and self-study.

Many colleges and universities offer formal *GIS graduate degree programs* (master's or certificate). This type of learning is recommended for those who want a thorough understanding of the principles of GIS.

Instructor-led classroom training is offered by GIS vendors such as ESRI or their partners. Instructor-led classes allow one to focus completely on the task at hand, free of daily distractions; they allow better instructor–student interaction. More information about training offered can be found on the software vendor's Website.

Web-based training lets one control the pace of learning and is very affordable. It works best for motivated, independent learners who want or need flexibility in their training schedule or location and for those who cannot afford the time away from work to attend traditional classes. Examples of institutions that offer Web-based GIS courses are ESRI's Virtual Campus, Penn State, Simon Fraser University, and the University of Southern California.

GIS Conferences abound. ESRI, for instance, runs the largest GIS conference in the world, the ESRI User Group Conference. In 2003, ESRI concluded its twenty-third annual user conference during which more than ten thousand GIS professionals gathered to share their experiences and knowledge. Papers presented at this and past conferences can be browsed at http://www.esri.com/library/userconf/archive.html. In 2001, ESRI launched an annual conference for users in education, which showcases applications in both K-12 and higher education.

Self-study is an option for anyone using GIS only casually, learning on one's own using workbooks and CD-ROMs. This type of learning works best for highly motivated learners who can do the work and master the content with little outside support. The books discussed earlier in this chapter constitute a good collection with which to start.

Daniel Teodorescu is the director of institutional research at Emory University in Atlanta.

INDEX

Back Issue/Subscription Order Form

Copy or detach and send to:

Jossey-Bass, A Wiley Imprint, 989 Market Street, San Francisco CA 94103-1741

Call or fax toll-free: Phone 888-378-2537 6:30AM – 3PM PST; Fax 888-481-2665

Back Issues: Please send me the following issues at $29 each
(Important: please include ISBN number with your order.)

_____

_____

_____

$ _____   Total for single issues

$ _____   SHIPPING CHARGES: SURFACE   Domestic Canadian

	First Item	$5.00	$6.00
	Each Add'l Item	$3.00	$1.50

For next-day and second-day delivery rates, call the number listed above.

Subscriptions   Please __ start __ renew my subscription to *New Directions for Institutional Research* for the year 2_____at the following rate:

U.S.	__ Individual $80	__ Institutional $150
Canada	__ Individual $80	__ Institutional $190
All Others	__ Individual $104	__ Institutional $224
Online Subscription		__ Institutional $150

**For more information about online subscriptions visit
www.interscience.wiley.com**

$ _____   Total single issues and subscriptions (Add appropriate sales tax for your state for single issue orders. No sales tax for U.S. subscriptions. Canadian residents, add GST for subscriptions and single issues.)

__Payment enclosed (U.S. check or money order only)

__VISA __ MC __ AmEx __ # _____ Exp. Date _____

Signature _____ Day Phone _____

__ Bill Me (U.S. institutional orders only. Purchase order required.)

Purchase order # _____
Federal Tax ID13559302   GST 89102 8052

Name _____

Address _____

_____

Phone _____ E-mail _____

For more information about Jossey-Bass, visit our Web site at www.josseybass.com

IR119   **Maximizing Revenue in Higher Education**
F. King Alexander, Ronald G. Enrenberg
This volume presents edited versions of some of the best articles from a forum on institutional revenue generation sponsored by the Cornell Higher Education Research Institute. The chapters provide different perspectives on revenue generation and how institutions are struggling to find an appropriate balance between meeting public expectations and maximizing private market forces. The insights provided about options and alternatives will enable campus leaders, institutional researchers, and policymakers to better understand evolving patterns in public and private revenue reliance.
ISBN: 0-7879-7221-5

IR118   **Studying Diverse Institutions: Contexts, Challenges, and Considerations**
M. Christopher Brown II, Jason E. Lane
This volume examines the contextual and methodological issues pertaining to studying diverse institutions (including women's colleges, tribal colleges, and military academies), and provides effective and useful approaches for higher education administrators, institutional researchers and planners, policymakers, and faculty seeking to better understand students in postsecondary education. It also offers guidelines to asking the right research questions, employing the appropriate research design and methods, and analyzing the data with respect to the unique institutional contexts.
ISBN: 0-7879-6990-7

IR117   **Unresolved Issues in Conducting Salary-Equity Studies**
Robert K. Toutkoushian
Chapters discuss the issues surrounding how to use faculty rank, seniority, and experience as control variables in salary-equity studies. Contributors review the challenges of conducting a salary-equity study for nonfaculty administrators and staff—who constitute the majority of employees, even in academic institutions—and examine the advantages and disadvantages of using hierarchical linear modeling to measure pay equity. They present a case-study approach to illustrate the political and practical challenges that researchers often face when conducting a salary-equity study for an institution. This is a companion volume to Conducting Salary-Equity Studies: Alternative Approaches to Research (IR115).
ISBN: 0-7879-6863-3

IR116   **Reporting Higher Education Results: Missing Links in the Performance Chain**
Joseph C. Burke, Henrick P. Minassians
The authors review performance reporting's coverage, content, and customers, they examine in depth the reporting indicators, types, and policy concerns, and they compare them among different states' reports. They highlight weaknesses in our current performance reporting—such as a lack of comparable indicators for assessing the quality of undergraduate education—and make recommendations about how to best use and improve performance information.
ISBN: 0-7879-6336-4

Get Online Access to
New Directions for Institutional Research

New Directions for Institutional Research is available through Wiley InterScience, the dynamic online content service from John Wiley & Sons. Visit our Web site and enjoy a range of extremely useful features:

WILEY INTERSCIENCE ALERTS
    Content Alerts: Receive tables of contents alerts via e-mail as soon as a new issue is online.
    Profiled Alerts: Set up your own e-mail alerts based on personal queries, keywords, and other parameters.

QUICK AND POWERFUL SEARCHING
    Browse and Search functions are designed to lead you to the information you need quickly and easily. Whether searching by title, keyword, or author, your results will point directly to the journal article, book chapter, encyclopedia entry or database you seek.

PERSONAL HOME PAGE
    Store and manage Wiley InterScience Alerts, searches, and links to key journals and articles.

MOBILEEDITION™
    Download table of contents and abstracts to your PDA every time you synchronize.

CROSSREF®
    Move seamlessly from a reference in a journal article to the cited journal articles, which may be located on a different server and published by a different publisher.

LINKS
    Navigate to and from indexing and abstracting services.

For more information about online access, please contact us at: North, Central, and South America: 1-800-511-3989, uscs-wis@wiley.com
All other regions: (+44) (0) 1243-843-345, cs-wis@wiley.co.uk

WILEY
InterScience®
www.interscience.wiley.com
Discover something great

www.interscience.wiley.com

United States Postal Service

Statement of Ownership, Management, and Circulation

1. Publication Title	2. Publication Number	3. Filing Date
New Directions For Institutional Research	0 2 7 1 - 0 5 7 9	9/30/03

4. Issue Frequency	5. Number of Issues Published Annually	6. Annual Subscription Price
Quarterly	4	$80 Individul / $150 Institution

7. Complete Mailing Address of Known Office of Publication (Not printer) (Street, city, county, state, and ZIP+4)

989 Market Street
San Francisco, CA 94103-1741
San Francisco County

Contact Person
Joe Schuman

Telephone
415 782 3232

8. Complete Mailing Address of Headquarters or General Business Office of Publisher (Not printer)

Same as above

9. Full Names and Complete Mailing Addresses of Publisher, Editor, and Managing Editor (Do not leave blank)

Publisher (Name and complete mailing address)

Wiley, San Francisco
Jossey-Bass - Pfeiffer
Address - same as above

Editor (Name and complete mailing address)

J. Fredericks Volkwein
Penn State University/Center For Study of Higher
400 Rackley Bldg.
University Park, PA 16801-5252

Managing Editor (Name and complete mailing address)

None

10. Owner (Do not leave blank. If the publication is owned by a corporation, give the name and address of the corporation immediately followed by the names and addresses of all stockholders owning or holding 1 percent or more of the total amount of stock. If not owned by a corporation, give the names and addresses of the individual owners. If owned by a partnership or other unincorporated firm, give its name and address as well as those of each individual owner. If the publication is published by a nonprofit organization, give its name and address.)

Full Name	Complete Mailing Address
John Wiley & Sons Inc.	111 River Street Hoboken, NJ 07030
Same as above	Same as above

11. Known Bondholders, Mortgagees, and Other Security Holders Owning or Holding 1 Percent or More of Total Amount of Bonds, Mortgages, or Other Securities. If none, check box ► ☐ None

Full Name	Complete Mailing Address
Same as above	Same as above

12. Tax Status (For completion by nonprofit organizations authorized to mail at nonprofit rates) (Check one)
The purpose, function, and nonprofit status of this organization and the exempt status for federal income tax purposes:
☐ Has Not Changed During Preceding 12 Months
☐ Has Changed During Preceding 12 Months (Publisher must submit explanation of change with this statement)

PS Form 3526, October 1999   (See Instructions on Reverse)

---

13. Publication Title	14. Issue Date for Circulation Data Below
New Directions For Institutional Research	Summer 2003

15. Extent and Nature of Circulation		Average No. Copies Each Issue During Preceding 12 Months	No. Copies of Single Issue Published Nearest to Filing Date
a. Total Number of Copies (Net press run)		1,831	1,962
b. Paid and/or Requested Circulation	(1) Paid/Requested Outside-County Mail Subscriptions Stated on Form 3541. (Include advertiser's proof and exchange copies)	589	525
	(2) Paid In-County Subscriptions Stated on Form 3541 (Include advertiser's proof and exchange copies)	0	0
	(3) Sales Through Dealers and Carriers, Street Vendors, Counter Sales, and Other Non-USPS Paid Distribution	184	89
	(4) Other Classes Mailed Through the USPS	0	0
c. Total Paid and/or Requested Circulation (Sum of 15b. (1), (2),(3),and (4))	►	773	614
d. Free Distribution by Mail (Samples, compliment ary, and other free)	(1) Outside-County as Stated on Form 3541	0	0
	(2) In-County as Stated on Form 3541	0	0
	(3) Other Classes Mailed Through the USPS	1	1
e. Free Distribution Outside the Mail (Carriers or other means)		81	82
f. Total Free Distribution (Sum of 15d. and 15e.)	►	82	83
g. Total Distribution (Sum of 15c. and 15f)	►	857	697
h. Copies not Distributed		974	1,265
i. Total (Sum of 15g. and h.)	►	1,831	1,962
j. Percent Paid and/or Requested Circulation (15c. divided by 15g. times 100)		90%	88%

16. Publication of Statement of Ownership
☐ Publication required. Will be printed in the Winter 2003 issue of this publication.   ☐ Publication not required.

17. Signature and Title of Editor, Publisher, Business Manager, or Owner

Susan E. Lewis   Date 9/30/03
VP & Publisher - Periodicals

I certify that all information furnished on this form is true and complete. I understand that anyone who furnishes false or misleading information on this form or who omits material or information requested on the form may be subject to criminal sanctions (including fines and imprisonment) and/or civil sanctions (including civil penalties).

Instructions to Publishers

1. Complete and file one copy of this form with your postmaster annually on or before October 1. Keep a copy of the completed form for your records.

2. In cases where the stockholder or security holder is a trustee, include in items 10 and 11 the name of the person or corporation for whom the trustee is acting. Also include the names and addresses of individuals who are stockholders who own or hold 1 percent or more of the total amount of bonds, mortgages, or other securities of the publishing corporation. In item 11, if none, check the box. Use blank sheets if more space is required.

3. Be sure to furnish all circulation information called for in item 15. Free circulation must be shown in items 15d, e, and f.

4. Item 15h., Copies not Distributed, must include (1) newsstand copies originally stated on Form 3541, and returned to the publisher, (2) estimated returns from news agents, and (3), copies for office use, leftovers, spoiled, and all other copies not distributed.

5. If the publication had Periodicals authorization as a general or requester publication, this Statement of Ownership, Management, and Circulation must be published; it must be printed in any issue in October or, if the publication is not published during October, the first issue printed after October.

6. In item 16, indicate the date of the issue in which this Statement of Ownership will be published.

7. Item 17 must be signed.

Failure to file or publish a statement of ownership may lead to suspension of Periodicals authorization.

PS Form 3526, October 1999 (Reverse)